UNIVERSITY OF STRATHCLYDE

30125 00412349 2

D1756368

Books are to be returned on or before
the last date below.

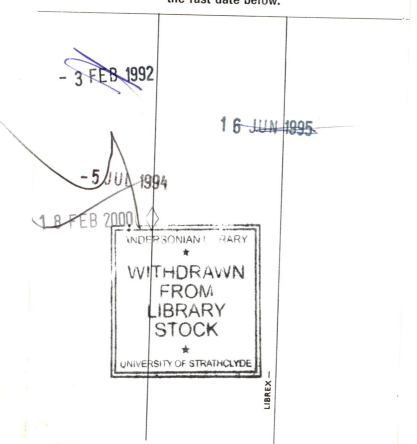

- 3 FEB 1992

1 6 JUN 1995

- 5 JUN 1994

1 8 FEB 2000

ANDERSONIAN LIBRARY
★
WITHDRAWN
FROM
LIBRARY
STOCK
★
UNIVERSITY OF STRATHCLYDE

LIBREX —

# Automatic Test Equipment

£24479239

# Automatic Test Equipment

*Keith Brindley*

D
681
BRI

ANDER⋯⋯  ⋯⋯RN

2 0. SEPT 91

UNIVERSITY OF STRATHCLYDE

Newnes
An imprint of Butterworth-Heinemann Ltd
Linacre House, Jordan Hill, Oxford OX2 8DP

PART OF REED INTERNATIONAL BOOKS

OXFORD   LONDON   BOSTON
MUNICH   NEW DELHI   SINGAPORE   SYDNEY
TOKYO   TORONTO   WELLINGTON

First published 1991

© Keith Brindley 1991

All rights reserved. No part of this publication
may be reproduced in any material form (including
photocopying or storing in any medium by electronic
means and whether or not transiently or incidentally
to some other use of this publication) without the
written permission of the copyright holder except in
accordance with the provisions of the Copyright,
Designs and Patents Act 1988 or under the terms of a
licence issued by the Copyright Licensing Agency Ltd,
90 Tottenham Court Road, London, England W1P 9HE.
Applications for the copyright holder's written permission
to reproduce any part of this publication should be addressed
to the publishers.

**British Library Cataloguing in Publication Data**
Brindley, Keith
    Automatic test equipment.
    I. Title
    681

ISBN 0 7506 0130 2

Typeset by Vision Typesetting, Manchester
Printed and bound in Great Britain by Billings & Sons of Worcester

# Contents

# Preface

First, an explanation of terms. Often, writers use terms such as *system under test* (SUT), *board under test* (BUT), *device under test* (DUT) or *unit under test* (UUT) to describe both appliances being tested and measurements taken. This is rather indiscriminate and ambiguous. As it is a specific function of an appliance which is always measured by test equipment – and not the appliance itself (you can find its height, width, depth and so on in the appliance specification!), I refer to the quantity being measured as the **measurand**: a term regularly used in the field of study of electronic instrumentation. I refer to tested systems, printed circuit boards or devices individually.

Electronic test equipment has come a long way since the days of meters and basic oscilloscopes. It used to be that testing of appliances was just a case of measuring a few independent analog measurand parameters such as voltage and current amplitude, frequency and time relationships and so on, at a small number of points. Generally, parameters could be measured one at a time, without problems.

Typical modern appliances, on the other hand, are of a microprocessor-based system nature and demand testing of a large number of digital and analog parameters at a correspondingly large number of points. Additionally, parameters are often so interdependent their values only have significance when monitored in relation with each other. Thus, measurements must be taken simultaneously and in real-time.

Trends in electronic test equipment naturally reflect this change and Figure 1 illustrates the general move from single-time, single-measured test instruments to multi-time, multi-measurand instruments. Simple analog and digital meters represent basic equipment, capable of performing a single measurement at a single time. Oscilloscopes extend measurements by performing them over a period of time. Dual- and four-trace oscilloscopes allow a small number of measurements to be made over this period. Logic analysers take this facility two stages further: first, by

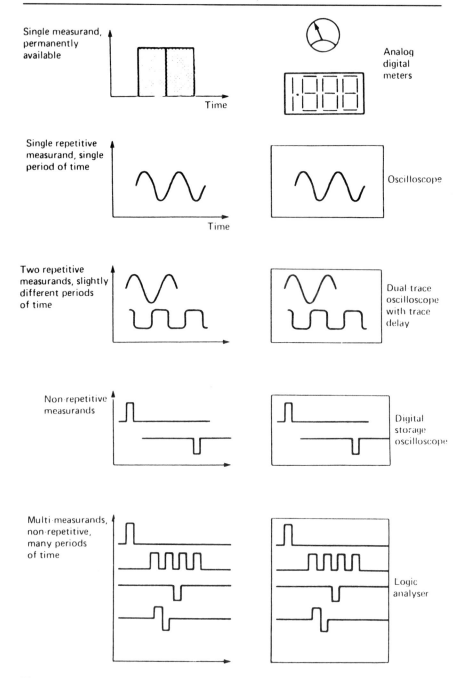

**Figure 1** *Trends in modern electronic test equipment correspond to a general move from single-time, single-measured instruments to multi-time, multi-measured instruments*

allowing a large number of measurements to be made over the period; and second, with recent developments in logic analysis, by allowing a large number of measurements to be made over a number of time periods.

In all these instances, however, a user effectively controls test equipment functions. Consequently, limits suggested here are not, in fact, test equipment limitations but human limitations. It simply becomes increasingly difficult to correlate all the information regarding the many measurements modern test equipment is capable of taking and, in many instances, is even impossible. Where test functions still remain humanly possible test procedures often take so long as to be uneconomic.

Which brings us to automatic test equipment (ATE); capable of doing all the measurements and tests we require, then presenting test results in a requested format – quickly and economically.

My main purpose in writing this book is to demystify automatic test equipment. Existing literature on automatic test equipment is not often written in a clear manner. There are reasons for this – the people who write about automatic test equipment must be the people who understand the topic; and these are usually engineers who look at automatic test equipment from inside-out. However, engineers are not renowned for high qualities of authorship. Other main sources of literature are worldwide standards. These, though, do not *explain* automatic test equipment; instead they merely formalize its component parts. From any existing literature therefore it is impossible to ask the question, *'What is automatic test equipment?'* and find a satisfactory answer. Automatic test equipment is a mystery simply because of this.

Tackling the situation from outside-in, on the other hand, as a writer (with engineering qualifications) trained as a technical author and journalist, I hope I have been able to explain concepts and standards in a much clearer way, passing on my understanding of the topic more successfully. This I hope I have been able to do without shirking technical considerations.

This book is for anyone who has an interest in automatic test equipment. Anyone in the industry who needs to know what types of equipment are available; what each type is capable of doing, what tests are performed, what computer buses are used, what the buses are capable of, and so on will find answers here. Managers, engineers, technicians, scientists, students, teachers, graduates, those in purchasing positions will benefit. It is a general-purpose book, which explains concepts: but is a reference book too, which defines specifications.

I have attempted to organize the book in a reasonably logical manner. My intention in this respect is to allow readers of whatever technical ability and specific knowledge to be able to use it. Chapters describe in successively greater details aspects covered, more generally, in earlier chapters. Consequently, it's a good idea to read from the beginning through to the end. However, that's not to say specific details cannot be accessed immediately, by turning straight to whichever chapter is required. Further, a glossary of important terms and considerations of automatic test equipment systems is included. This is, of necessity, a fairly involved glossary. After all, automatic test equipment is quite involved, itself. Names and addresses of important organizations in automatic test equipment are included, as is a list of books, articles and papers for suggested further reading.

*Keith Brindley*

# 1 What is automatic test equipment?

Automatic test equipment comes in many forms. Indeed, it is difficult to define *precisely* what the term automatic test equipment means. Lowest common denominator suggests automatic test equipment to be any item of test equipment controlled by a computer. Yet this, in itself, is not totally complete because some items of test equipment have an internal computer (in microprocessor form) to monitor and control certain functions automatically, making the equipment easier to use.

Thus, an instrument such as a spectrum analyser could feature automatic control of functions such as frequency scan, centre frequency and resolution bandwidth: it is an example of an **automatic test instrument**. It is not necessarily, however, what we class as automatic test equipment.

Most modern automatic test instruments are **programmable**; they feature an interface allowing their internal microprocessor (and hence their measurement functions) to be controlled by another computer or microprocessor. Usually most, if not all, measurement facilities of an automatic test instrument may be set by a computer via this interface, and measurements taken are similarly relayed back to the computer for correlation and display. When a computer is used to control one or more programmable automatic test instruments the resultant system is what we know as **automatic test equipment** (ATE). This difference is important: automatic test instruments are devices capable of performing and displaying measurements autonomously or in a system; automatic test equipment is a complete measurement system, consisting of one or more automatic test instruments and a computer controller.

Automatic test equipment requires computer control to ensure correct operation, record measurements, correlate vast amounts of measurement data, and present data in a form understandable by human users. In effect, users no longer *directly* control automatic test equipment (although users must still program the computer which *does* control it) and most, if not all, functions are automatic.

Measurements are not limited by users: any number of measurements can be performed in any number of time periods. For example, the user of an analog voltmeter has great difficulty in taking and recording even one measurement a second. Programmed automatic test equipment may take, record and display as many as, say, one thousand measurements in the same time. Alternatively, automatic test equipment may take and record one measurement every second for the next thousand days – non-stop and accurately (without food, drink or sleep!).

## Types of automatic test equipment

There is a large number of types of automatic test equipment. These types are defined, basically, by the way they set about testing products. Is power applied to the product? Are external inputs applied, as if the product were in its real-life application? Can individual components within the product be tested, in isolation from all other components? It doesn't take many questions like these to show there are almost as many pure types of automatic test equipment systems as there are manufacturers of automatic test equipment systems. After all, each manufacturer likes to include at least a few features which define its automatic test equipment system as being different, better, cheaper, than the rest.

Fortunately, this large number of types can be generalized into only a handful of categories. These are described in Chapter 2.

## Fixtures

Any automatic test equipment system must be connected to the product it is to test. How this connection comes about is usually a matter of electromechanical interfacing; via connectors, probes and so on. Those parts of an automatic test equipment system used to interface to products are given the name fixtures and some basic types exist. These basic fixture types, and some possible future types, are discussed in Chapter 3.

## Test strategies

Obviously, the earlier definition of automatic test equipment: a programmable, computer-controlled system, is a basic one. It leads you to reason that such a system is all-encompassing and may, simply by changing the program, be used to perform any measurement task. This is not, of course, the case; you would not use top-of-the-range automatic test equipment costing, perhaps, £1 million to measure resistance of a resistor. That is the test equipment equivalent of using a sledgehammer to crack a nut.

In the other extreme you would not expect a cable harness checker to be able to tell you what is wrong with the mother-board of a malfunctioning nuclear-tipped missile guidance system. Although in the story David won, it would take an even greater fluke to beat Goliath, here.

These two examples serve to illustrate the need for an effective **test strategy**, which carefully plans and defines an organization's requirements for automatic test equipment. A test strategy can not only help an organization to choose the correct type of automatic test equipment for its technical purposes, but also shows the most economically suited solution. This may be a considerable advantage, where equipment can be as cheap as just a few thousands of £s up to a few millions. Test strategies, costs and, indeed, even the reasons for need of automatic test equipment are discussed in Chapter 4.

## Test methods and processes

Part of this test strategy an organization must have, is a definition of individual tests which must be performed on products, and a consequent understanding of processes which those tests must follow. Definition of required tests depends almost totally on the products to be tested. Products should be assessed in quality terms: likelihood of failures; where failures are likely to occur; abilities of personnel involved in design to make products easier to test, and so on. Factors such as these have direct relevance to the test processes and, hence, test strategy used. Test methods and processes are discussed in Chapter 5.

## Basic methods of creating automatic test equipment

There are three basic methods of creating automatic test equipment. First, a unique system may be designed and made, specifically for the purpose. Figure 1.1 shows an example, capable of taking a number of measurements of voltage and current, while counting events, measuring frequency, distortion and frequency response, and monitoring signals on a data bus. Output from the device to the appliance being measured is a swept sine-wave signal. Control of the various measuring facilities is provided by the microprocessor-controlled heart of the device, which in turn is controlled by programmed instructions from the user. This type of automatic test equipment is, in fact, a computer system complete with the necessary input and output units to allow measurements of the various measurand parameters of the appliance tested. Recording of the values of these measurand parameters and the format of the correlated information again depends on the user's programmed instructions, and is displayed either on a monitor or hard-copied onto paper with a printer.

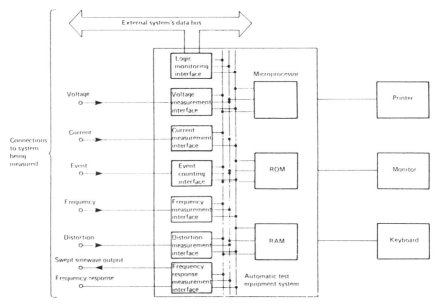

**Figure 1.1**   *Turnkey automatic test equipment system, in which equipment is a unique device, built for one application*

Such automatic test equipment is known as a **turnkey** system. It is a custom-built *device*, likely to be quite expensive in terms of initial capital outlay, can generally only be used to test one particular appliance and will probably be used to test electronic appliances (such as printed circuit assemblies) manufactured in vast quantities. In such a situation high capital outlay is justified against higher reliability it gives the manufactured appliances.

Second, microprocessor-based computers may be used to control general-purpose automatic test instruments such as meters, universal counter timers, logic analysers and signal generators, treating them as peripheral devices, as illustrated in Figure 1.2. In this method each peripheral test instrument performs measurements on the system being tested under central computer control, relaying readings to the computer which records and displays the correlated data onto a monitor or onto paper. As in the first method, the user controls overall system operation with programmed instructions.

This method gives a custom-built automatic test equipment *system*, still quite expensive but which can be adapted to allow its use in other test applications. Thus more than one particular appliance can be tested, so the system will most probably be used to test electronic appliances, say, manufactured in quite small quantities (with the knowledge that the system

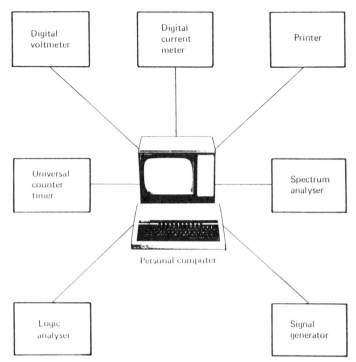

**Figure 1.2** *Rack and stack automatic test equipment system, in which a personal computer or any other microprocessor-based computer is used to control programmable readily-available test instruments*

can be easily adapted to suit other applications as and when necessary). Often an automatic test equipment system along these lines is called a **rack and stack** system.

In practice even the first method can usually be adapted to test more than one appliance, as a modular design approach is often used which allows the user (or at least the automatic test equipment supplier) to change measurement modules and software to suit other applications. These two methods are merely simplified representations of extremes of automatic test equipment design philosophy. In the first method, the range of test instruments is built into a complete computer-controlled device; in the second method the range is simply a collection of individual but interconnected instruments, controlled by a central computer.

If a number of instruments are to be controlled by a single, central computer it makes sense if each instrument's two-way interface to and from the computer links to common data and control buses. Connections between individual instruments and computer are simplified enormously – often wiring is simply a matter of linking instruments with ribbon cable, as

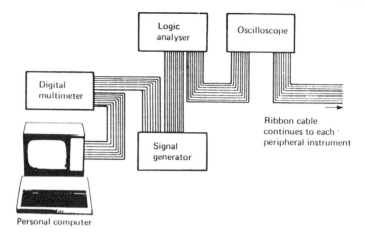

**Figure 1.3**   *Another method of creating a rack and stack automatic test equipment system, in which instruments and controller are connected with common data and control buses*

shown in Figure 1.3. Bus structures for automatic test equipment make corresponding systems very flexible as extra instruments may be added with little fuss, and changing system test requirements is simply a matter of changing instruments and reprogramming computer controllers. There are, needless to say, many bus standards in use, and main types are discussed in Chapter 6.

Rack and stack automatic test equipment systems' use of central computer controllers with peripheral instruments points the way to the third method of making automatic test equipment. Recently a trend has developed, building on the automatic test equipment bus principle, in which the bus is controlled not by a stand-alone computer but by a purpose-built, standard-sized computer controller. Further, peripheral instruments under computer control are also standard-sized and in modular form. A major benefit of this trend is that a single housing may then be used to hold all instrument modules. All functions and features of each peripheral instrument (including power supply) are controlled and adjusted by the computer via the buses.

Obviously system cost can be much lower (estimated at around one-third the price of an equivalent non-modular system) and overall size is considerably reduced (about one-tenth the equivalent non-modular system). A modular system ensures an automatic test equipment system is easily adaptable to future requirements: when a different system is needed, unwanted modules may be taken out of the housing and new modules simply slotted in.

Another important benefit is the extremely high data rates which may be allowed with a purpose-built bus, a factor important where a large automatic test equipment system, with many measurands and programmed steps, is required.

For want of a better generic name, such automatic test equipment systems may be called **modular** instrument bus systems.

On a broad outlook, the method used to make automatic test equipment is irrelevant. Although each has its advantage and disadvantages which guide prospective users to choose one particular method to suit a particular application, in the end each method of construction is there to do but one job – to test manufactured products. The method used to make the automatic test equipment needed to test the products a user manufactures is simply a means to that end.

### A comparison of automatic test equipment system types

In reality, of course, all automatic test equipment systems constructed by whatever method are organized around computer-controlled instrument bus systems. Table 1.1 is a comparison of all three methods, giving details about each part of each system type.

Such is the importance of the three methods of making automatic test equipment, however, now and in the foreseeable future, that the last three chapters in this book discuss specific ways these three methods are usually (not always!) organized – detailing important characteristics and attributes. The usual way rack-and-stack automatic test equipment systems are organized is with an interface bus known as the **general-purpose interface bus** (GPIB), which is discussed in Chapter 7. A common interface bus used to organize turnkey automatic test equipment systems is **VMEbus**, discussed in Chapter 8. Modular automatic test equipment systems are most often organized around the **VXIbus** interface bus, discussed in Chapter 9.

## Advantages of automatic test equipment systems

In many instances, use of automatic test equipment systems is a foregone conclusion. Where manufactured appliances are complex, human testing ability is stretched beyond its possibilities. Here the advantages of automatic test equipment systems are taken as a matter of course:

- More measurements – all measurements in an automatic test equipment are performed automatically, with a consequent increase of speed, so it is possible to increase the number of measurements made.
- Greater accuracy – many errors may be introduced in manual test equipment systems, often because measurements are taken once, at a

**Table 1.1**  Comparing automatic test equipment systems of three main types

| Automatic test equipment method | Instruments | Bus | Controller | System |
|---|---|---|---|---|
| Turnkey | Often purpose-built<br>May be modular | Often specific to system<br>May be a standard bus | Specific to system | Custom-built device |
| Rack and stack | Individual | Standard bus (e.g. GPIB)<br>May be specific | Often a personal<br>computer | Custom-built system,<br>adapted devices |
| Modular | Modular<br>May be purpose-built | Standard bus (e.g.<br>VXIbus) | Often specific to system<br>May be personal<br>computer | Modular, off-the-shelf,<br>integrated |

single time. Automatic test equipment systems may offer a solution by taking multiple readings and averaging results; errors are reduced and may even be eliminated.

• Faster procedures – where measurements entail complicated setting up, triggering, result interrogation and evaluation procedures manual measurement systems are inevitably slow. Human involvement means these processes must be undertaken at human speed; automatic systems, on the other hand, can perform the same procedures much faster.

• Elimination of human involvement – automatic procedures avoid human interpolation of results, subjective result interrogation, subjective evaluation and erroneous recordings of measurements.

Often, high expense of automatic test equipment may be offset against greater reliability of appliances tested. An amortization time of less than two years is common. After that period the automatic test equipment system, far from being an expenditure actually makes money. Customer benefits of greater reliability and cheaper products are well documented.

There is no doubt: use of automatic test equipment, in itself, is an advantage – an advantage manufacturers must be careful not to miss out on (1) by not using automatic test equipment (2) by not choosing the right type for the application, in the first place.

# 2 Types of automatic test equipment

Automatic test equipment is available in many forms. Part of its nature is its chameleon-like ability to test whatever is testable. Automatic test equipment is, as we have seen in Chapter 1, computer-based equipment and as such is governed by the computer principle of deferred design – simply by giving the computer a suitable program and providing it with necessary input and output peripherals it will do just about anything. Thus, in one example, automatic test equipment is found doing the simplest of tasks such as measuring components' resistances while, in another, it is capable of testing the most complex printed circuit assemblies; pin-pointing defective components to make repair easier.

Complexity of automatic test equipment, on the other hand, compared with individual test instruments such as meters, oscilloscopes and logic analysers, is such that it is possible to itemize particular areas of involvement requiring specific equipment. Contrasted with individual test instruments, where an oscilloscope is an oscilloscope whether it's used on the bench or on the production line, automatic test equipment is more likely to be purpose-built for one application.

Generally, applications where automatic test equipment is used to test a product parallel the normal stages of that product's life. Thus, automatic test equipment systems may be used in a product's:

- Design and development.
- Production.
- Reliability and certification test.
- Service.

More often than not, different automatic test equipment systems are used at each stage, although it is possible to design a single system with the capability to test the product at every stage.

There is a number of types of automatic test equipment, broadly categorized into the main tests performed. There is, however, no reason why more than one type of test cannot be performed by a single system. Similarly, there is no reason why a single system cannot perform *all* tests required: indeed, the trend is towards this. Use of a computer-controlled bus with modular peripheral instruments, as discussed here and in later chapters, makes such an encompassing automatic test equipment system possible.

Main categories of automatic test equipment include:

- Component testers – to test individual parts prior to assembly within a product to ensure they fall within specified tolerances.
- Unpackaged assembly testers – to test assembled parts, say, printed circuits, prior to packaging. This is the main area of concern for this book.
- Packaged assembly testers – to test and ensure reliability of the complete and packaged product prior to use by a customer.
- Maintenance and service equipment – to repair and overhaul a used product.

These are all very broad categories, though, and we must consider each in detail. One or more categories of test may be used in any particular application, and those categories used depend on an organization's overall test strategy (see Chapter 4).

## Component parts test equipment

Testing of component parts is rarely performed in anything other than the manufacturing stages of a product. Nevertheless considerable test equipment is devoted to the testing of component parts and so must be considered.

Component parts testing is a fairly basic procedure. Generally, it is accomplished using simple procedures designed to determine measurands such as resistance, capacitance, semiconductor functions, dimensions, solderability and continuity. Testing is performed simply to ascertain the parts are of a specified quality. Tests inevitably depend on the parts.

Passive component parts are generally tested simply to measure their value and dimensions. Simple meter or bridge circuits can be used to perform such tests manually or semi-automatically. More complex, automatic measurements may be made using instruments which incorporate bridges, analog-to-digital conversion, function generators, voltage supplies, analog and digital stimuli and so on.

Active components are usually tested on a functional basis, that is, they are supplied with stimuli which simulate operating conditions, and resultant relationships are measured and compared with the ideal. Again, this may be a manual, semi-automatic or automatic test.

**Photo 2.1**  *Close-up of APT-2200N fixtureless tester, using moving probes to test unassembled printed circuit board (Contax)*

Printed circuit boards are often tested prior to assembly. Such **bare-boards** can be of a range of complexity; from simple single-sided boards which will hold only a handful of through–hole (that is, leaded) components, to extremely complex multi-layered boards (perhaps with over thirty internal track layers) designed to hold hundreds of surface mounted (that is, leadless) components. Consequently, test equipment used varies largely according to the circuit boards to be tested.

Simple boards may require just a straightforward visual check, perhaps with a magnifying aid. Visual checks are not reliable on more complex boards, though, and so test equipment which checks continuity of board track is common. A simultaneous check of insulation between tracks is recommended too. Automatic continuity and insulation test equipment – in which the bare-board is placed on a bed–of–nails fixture or, featuring a pair of moving probes programmed to position at large numbers of test points around the board – is available. With multi-layered boards it is often impossible, though, to check internal layers for continuity.

Automatic optical inspection (AOI) using cameras, scanning lasers, or sometimes X-rays is used to compare bare-board tracks with an ideal image (often called the **golden image**). Such systems, however sophisticated, cannot absolutely guarantee continuity or insulation.

Similar to the requirements of continuity and insulation testing in circuit boards, cable harnesses and backplanes of complex multi-board products need to be checked prior to assembly. Similar continuity and insulation checking test equipment is therefore available, too. It is becoming increasingly popular to use time domain reflectometers to provide a graphic signature of the harness or backplane, indicating presence and type of fault or similarity to a golden signature.

There is a method which sidesteps requirement for testing component parts of a product – to use parts which are known and guaranteed to be of required quality. National, regional and international standards organizations have coordinated (and continue to do so) standards and procedures which ensure components manufactured by a supplier are of a defined quality. Standards and procedures effectively form complete specification systems, incorporating all types of components and manufacturers.

BS9000 is the British specification system, CECC system operates in the UK and Europe, while IECQ system operates in the UK, Europe and worldwide. Approved components are listed in frequently updated qualified products lists (QPLs) and it is a simple matter for the purchaser to identify the required components from these lists prior to purchasing.

This sort of self-assessment procedure, known commonly as **vendor assessment**, can aid product manufacture enormously and is all part of commonly accepted methods of improving quality. Accurate manufacturing times can be predicted and, ●verall, considerable wasted time may be eliminated. When component parts are purchased outside such a procedure, **purchaser assessment** remains the only viable method of assuring a finished product's quality. Fortunately, test equipment to carry this out is easily available and not too expensive.

Readers are referred to another of the author's books; *Newnes Electronics Assembly Handbook*, for a detailed discussion of the subject of quality, standardization and specifications systems.

## Unpackaged assembly test equipment

Once assembled and soldered, but prior to packaging, it is usual to test printed circuit boards to ensure the complete assembly performs as expected. Test equipment to do this could be used in maintenance and service environments, too, but is usually restricted to manufacturing stages simply because of large size. As far as this book is concerned, it is this area of automatic test equipment which is most important.

Three main categories of test (although other relatively minor categories exist) may be undertaken on unpackaged assemblies: in–circuit; functional; combinational, each of which has sub-categories.

### In-circuit test

Performed by accessing nodal points within an assembly, generally with the use of a bed-of-nails fixture, then testing individual parts of the circuit,

**Photo 2.2** *Zehntel 875 in-circuit test system (Teradyne)*

often to a component level. Often, bed-of-nails fixtures have many hundreds, if not thousands, of test probes. By comparing the measured values with defined ideal values, faults such as short circuits, misplaced components, wrongly valued components, poor soldered joints and defective tracks can be isolated. Tests on components are performed sequentially so that, depending on circuit complexity and numbers of components, a complete procedure may take considerable time. Nevertheless, it has been reported that some 90% of manufacturing faults can be detected by in-circuit testing, so it can represent an extremely powerful procedure. However, it is not capable of testing overall performance, in a real-time dynamic situation.

## Manufacturing defects analysis

Inherent in the use of in-circuit testing is the assumption that most problems occur in the manufacturing stage – in other words they are manufacturing defects. For this reason certain in-circuit testing techniques are known as **manufacturing defects analysis**, although this term strictly refers to in-circuit testing when no power is applied to assembly, testing the assembly passively. Manufacturing defects analysers are correspondingly cheaper than pure in-circuit testers, though they cannot check individual active devices for correct function. Another name often given to in-circuit testing by manufacturing defects analysis is **pre-screening**.

## Functional test

Referring to a test procedure which involves application of power and test signal patterns which simulate normal operation of the assembly, allowing a check of overall performance to be made. In this respect, functional testing is usually more rapid than in-circuit testing, and is often a simple pass or fail test (sometimes called **go/no go** testing) which can be of benefit in high volume production of assemblies. On the other hand, little indication of a specific fault is given in such a straightforward test, compared with in-circuit testing. Consequently, further tests may have to be undertaken to isolate a specific fault. These further tests may be performed on the same test equipment system, or on another system, and usually involve manual placement of probes in a manner defined by the program which controls the functional test system. Thus specific step-by-step tests can be undertaken, each of which depends on results of previous tests, until individual components or parts of an assembly are isolated as faulty. The ability of a functional test system to identify and isolate faults is thus dependent on the complexity of its test program.

Access to test nodes of the assembly is usually through the assembly's board connectors. This is something of a limiting factor as far as testing is concerned, because connectors only allow access to a small number of circuit nodes. Compared with the many hundreds of possible test probes available in bed-of-nails fixtures, used typically with in-circuit test equipment systems, functional test systems are at a disadvantage as they are not capable of testing individual components directly. However, good assembly designs (see Chapter 4), and some new test methods (see Chapter 5) are reducing this inherent disadvantage considerably.

## Combinational test

Neither in-circuit test nor functional test procedures alone allow all faults to be identified and isolated. In-circuit test cannot test an assembly in real-time – so functional faults cannot be detected. Similarly, functional test cannot directly isolate individual faulty components or parts of an assembly. Consequently, a recent test procedure combines both in-circuit and functional test procedures in one automatic test equipment system. Known, as you might expect, as **combinational test** these systems accentuate advantages of each individual procedure, while attempting to eliminate disadvantages.

**Photo 2.3** *Easytest 500AD polyfunctional test system (SPEA)*

At its most basic concept, combinational test undertakes each procedure – in-circuit and functional – separately, in a serial manner. However, this overlooks the real significance of combinational test's potential. Using various test methods and processes (see Chapter 5) combinational testers may electrically isolate parts of the whole assembly and perform both in-circuit and functional tests on those parts. By performing such tests in a logical and controlled manner, faults within each part may be isolated much more effectively than separate in-circuit and functional procedures allow.

Electrical isolation of parts of a circuit is known as **partitioning**. Such partitions are commonly called **clusters** (because they comprise clusters of individual components), so combinational test procedures of this category are often called **cluster tests**.

Where the potential of combinational test procedures is undoubtedly realized, however, is if size and position of such clusters within a circuit as well as tests carried out on each cluster may be decided upon by the combinational test equipment system – as it carries out its tests. Ability to do this is known variably as **multimode testing**, **polyfunctional testing**, or **variable in-circuit partitioning** (VIP), different manufacturers of such automatic test equipment systems choosing different names. Such a combinational test procedure potentially allows any partition allows any partition size from single components, to component clusters, to complete sub-circuits, through to entire assemblies to be tested according to requirements. This effectively allows a continuous capability from full in-circuit test to full functional test, defined automatically and dynamically by the automatic test equipment system as it tests individual assemblies. Interestingly, combinational test using these procedures is documented as reaching extremely high levels of fault detection and isolation. Indeed, over 99.5% (up to 100% in some cases) of all faults are reported to be detected and isolated by polyfunctional test procedures.

### Optical inspection

One other category of test used with unpackaged assemblies is optical inspection. It is not yet, however, of such significant importance to be discussed in depth, here. Optical inspection (particularly in an automated form) is possible using visible light, laser, infra-red or X-ray techniques discussed for bare-board testing. Correct placement of components, short circuits, open circuits, faulty soldered joints and so on can all be isolated using automated optical inspection.

## Packaged assembly test equipment

Once packaged in its housing a circuit board assembly may be subjected to tests to determine performance under stress conditions. These are generally of two types. First is a test procedure which simply determines whether the product operates under the stresses. Sometimes customer specifications (particularly of military origin) call for such stress tests and detail exact methods – these are known as **certification** *or* **qualification** tests.

### Screening tests

Second, perhaps of greater potential, are tests designed to force failures to occur prior to delivery to the customer. Such failures, which would otherwise occur during the product's working life (during the early cycle of a product's reliability curve) are forced to occur during the tests by abnormal stresses imposed, including in order of effectiveness (Institute of Environmental Sciences):

- Temperature cycling.
- Vibration.
- High temperature.
- Electrical stress.
- Thermal shock.
- Sine vibration, fixed frequency.
- Low temperature.
- Sine vibration, sweep frequency.

- Combined environment.
- Mechanical shock.
- Humidity.
- Acceleration.
- Altitude.

Tests such as these are known as **reliability screening tests**, sometimes **environmental stress screening tests**.


## Maintenance and service tests

Generally, maintenance and service tests are of a simpler nature than any of the tests detailed earlier. Service personnel can only take a limited amount of test equipment on-site, after all.

Tests which are used tend to be of a functional nature, simply to test products which have previously worked and for some reason have ceased to do so. Often these are returned as circuit board assemblies (replacing the faulty assembly with a known working assembly) to the manufacturing site, where it is known that complex in-circuit or functional test equipment, used to test assemblies on original manufacture, is located. Faults are in the main due to failure of single components, so in-circuit testing usually locates faults better than functional testing.

On-site, fairly basic test equipment prevents faults from being located rapidly, and time spent repairing faults under such conditions is not usually cost-effective; straightforward assembly replacement being a cheaper option.


## Automatic test equipment administration

Complexity of much automatic test equipment is such that day-to-day running of the system represents a significant proportion (estimated at around 50%) of total investment. Put another way, whatever is spent purchasing automatic test equipment can be expected to be spent again maintaining, programming, adapting the system, and training users.

Figure 2.1 is a pie chart showing approximate percentages of amounts as a proportion of total spent on automatic test equipment systems. This is, of course, based on typical results and actual amounts will vary, depending on the type of automatic test equipment system used in individual applications. From it, however, it is easy to see costs are not just dependent on the initial capital costs of automatic test equipment system purchase. There are other, ongoing, costs which must be considered. These ongoing costs increase with time, typically exceeding the initial capital cost of system

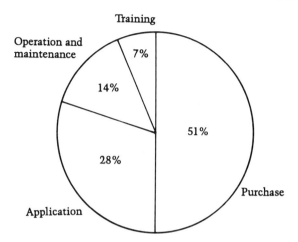

**Figure 2.1**   *Showing approximate percentages of amounts as a proportion of total cost of automatic test equipment systems for main overall costs*

purchase within one or two years; exact time depending on system type and application.

Ongoing costs may be divided into two basic groups – application costs, operating costs. Application costs include:

- Adapting the system to suit products to be tested.
- Programming the system to test products.

Operating costs include:

- Using the system – operators' labour costs.
- Reworking products found faulty – repair personnel labour costs.
- Retesting reworked products – operators' labour costs.
- Maintaining the system.
- Training of associated personnel.

## Adapting automatic test equipment systems

Fixtures (see Chapter 3) of some description are used to adapt an automatic test equipment system to suit particular assemblies. For example, in-circuit testers, combinational testers and some bare-board testers use bed-of-nails fixtures; while functional testers generally use printed circuit connectors.

A fixture is almost certainly assembly–specific, not usable for any other assembly-type. Consequently, design of a fixture is most likely to be done while the program is under generation, often by the programmer. A further consequence is that any minor changes in the assembly hardware must be matched by corresponding changes in the fixture.

## Programming automatic test equipment

Whatever type or complexity of automatic test equipment, a program is required to control it. Usually this is written in a high level computer language. Always the program is specific to the tested assembly-type – a new program is required to test a different assembly-type.

There are two basic methods of producing programs for automatic test equipment:

- Manufacturer of the automatic test equipment provides a program service (or recommends an outside software house) to generate programs. This has the disadvantage that the equipment manufacturer does not have an in-depth knowledge of the assembly to be tested.
- Manufacturer of the product to be tested generates programs. Disadvantage here being the assembly manufacturer does not have an in-depth knowledge of the automatic test equipment. On the other hand, if the automatic test equipment is to be re-programmed to test subsequent assemblies, knowledge gained from the generation of the first program is available.

Generally, program complexity – hence cost – depends on the test procedure used to test a product. Manufacturing defects analysers, for example, use relatively simple controlling programs which are consequently relatively cheap to write. In-circuit testers performs more complicated tests, so programs are more complex and more expensive. Functional testers perform relatively complex tests if they are to detect and isolate individual faults, so programs may be extremely complex and similarly expensive, depending on what is asked of the test equipment system.

Often manufacturers provide program development tools, typically in the form of component or routine libraries. Thus common test methods and processes for particular applications are available as stored routines, easily available when programs are being written. Generally, the longer a particular automatic test equipment system has been in use, the more of these routines are available.

## Maintenance of automatic test equipment

After an initial warranty period, it becomes the owner's responsibility to ensure equipment is adequately maintained. Maintenance falls into two main areas: service; calibration.

Service is often aided by self-test routines incorporated into the test equipment, allowing users to check equipment performance. In such equipment the most common method is to connect a self-test adaptor to the equipment's fixture, then run the routines. Labour costs of service personnel must be taken into account. Alternatively, test equipment manufacturers provide service support.

Calibration, where applicable, should be undertaken according to manufacturer's instructions. Sometimes calibration is automatic and no user involvement is required. Sometimes extra equipment is required to assist in calibration. Sometimes, equipment, or parts of it, must be returned to the manufacturer for factory calibration. Calibration personnel labour costs must not be forgotten.

### Training users of automatic test equipment systems

Three levels of staff are likely to come into contact with automatic test equipment:

- Operators.
- Engineers, for maintenance, design, development and production purposes.
- Programmers.

Each level of staff requires training to be able to perform adequately. The more complex the test equipment, the more training is likely to be necessary for all levels.

Normally, automatic test equipment manufacturers provide training courses, usually on-site although, where test equipment is particularly complex, training may be given at the supplier's base. An alternative solution, particularly where automatic test equipment systems are to be adapted to many products and so need to be re-programmed regularly, is to have in-house training personnel. In this case labour costs of such personnel must be taken into account.

## Type and availability of automatic test equipment

It is not the nature of this text to discuss individual examples of automatic test equipment. Readers are referred instead to manufacturers' data and to the ERA Technology report *Guide to low-cost ATE*, which details over 50 automatic test equipment systems costing under £50,000.

# 3 Fixtures

Between automatic test equipment and products (usually assembled and soldered printed circuit boards) some form of fixture needs to be used. Fixtures adapt any particular automatic test equipment system, allowing it to be used with any product. A standardized interface exists on any particular automatic test equipment system (often all a manufacturer's range of systems incorporate the same type of interface), which the device-specific fixture literally plugs into. If a new product is to be tested it is then a matter of changing the fixture – not the system.

Generally, fixtures are designed to allow tested products to be connected and disconnected from the automatic test equipment system rapidly and easily. Thus, after each individual product is tested it may be removed and the next product installed in the minimum of time. In high-volume production areas, mechanical, robotic handlers are often used to install printed circuit boards into fixtures, remove them, and install the following boards. Consequently one of the aims of fixture design must be to make installation and removal of tested boards as straightforward as possible.

Often, the type of fixture used in an application depends on the type of automatic test equipment system used. Thus automatic test equipment which requires internal access to nodes within the tested product's circuit requires a fixture which allows this. Manufacturing defects analysers, in-circuit testers, and combinational testers, therefore typically use fixtures with large numbers of probes, over which the product assembly is aligned then pressed onto. Each probe in the fixture is positioned at a point corresponding to a required circuit node, and as the board is pressed onto the fixture the probes make electrical contact.

For obvious reasons such fixtures are known as **bed-of-nails fixtures** and, typically, use spring-loaded probes which allow a measure of mechanical compliance while ensuring electrical contactability.

Functional testers, on the other hand, do not require access to internal circuit nodes, instead simply wishing to simulate a product's operational

**Photo 3.1** *Close-up of 5512 automated optical inspection head for assembled printed circuit board (Universal)*

environments in terms of power and interface signals. Consequently, fixtures on functional testers may be as simple as a printed circuit board plug connector, fitting into the corresponding socket connector on the product assembly.

Sometimes, **moving probe fixtures** are used to test unassembled or assembled circuit boards. These are effectively computer-controlled robots, with two or more arms, usually operating in an X–Y coordinate basis in the horizontal plane. Underneath each arm is located a probe. By aligning the tested product assembly underneath the arms and moving each arm to the coordinate position required, then lowering it vertically, probes are allowed to access circuit nodes within the product assembly.

All fixtures mentioned so far are electromechanical, involving making contact between connectors of one form or another and the product assembly. A number of non-contact probing methods are in early stages of development and use. These are likely to become important as products, generally, become more complex and dense in terms of numbers and complexities of components on products' boards.

## Bed-of-nails fixtures

Bed-of-nails fixtures are grouped according to actuation method – in other words, how the mechanism operates and ensures a tested board is fixed into its correct position, ready for test procedures to begin. Main groups are:

- Manual – boards are held in place by hand.
- Mechanical – boards are clamped in position. Suitable for boards with up to about 400 test point probes.
- Vacuum – an air pressure reduction on one side of the board holds it in position. Suitable for boards with up to about 20,000 test point probes.
- Pneumatic – pneumatic liquid at high pressure on one side of the board forces it into position. Suitable for boards with over about 20,000 test point probes.

Vacuum-actuated bed-of-nails fixtures are most common in test situations with single-sided and low- to medium-density printed circuit boards, allowing a wide range of numbers of test point probes to be accommodated in a fixture.

**Photo 3.2** *Factron vacuum fixture for Schlumberger 700 series testers (In-circuit Test)*

**Photo 3.3** *Testronics 101A low-cost fixturing system (DCA Technology)*

**Photo 3.4** *Factron manual test fixture, intended for low-volume production and prototype test (In-circuit Test)*

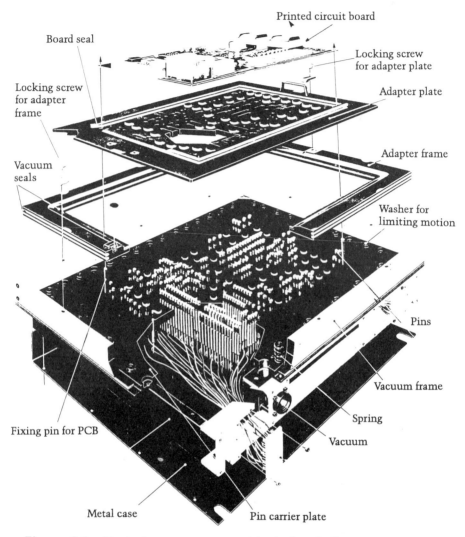

**Figure 3.1**   *Typical vacuum-actuated bed-of-nails fixture*

One problem with bed–of–nails fixtures is associated with numbers of test point probes. Each probe is spring-loaded, so more probes means greater pressure is required to hold the board in position. Consequently fixture designers must be aware of pressure requirements. Each probe has a spring resistance (see later) which is multiplied by the total number of probes to give total pressure required to hold the board in its test position. Actuation method must be able not only to supply this total pressure, but also to supply it uniformly over the board surface area to prevent bending, which may result in non-contact of some test point probes.

A typical vacuum-actuated fixture is shown in Figure 3.1. When a vacuum is applied to the vacuum port, atmospheric pressure above the fixture forces the board down onto the test point probes, uniformly with little or no board bending. No more than a few millimetres of movement from relaxed to test position is required.

Bed-of-nails fixtures have limitations in the number of test point probes available. Generally, test point probes are matrixed on fixed centres of typically 2.54 mm (0.1 inch), though sometimes a closer grid of 1.91 mm (0.075 inch) or 1.27 mm (0.05 inch) centres are used (occasionally centres of 0.63 mm – 0.025 inch – are used to provide fixtures for small substrates of extremely high component densities). This limitation is mainly due to pressure required to hold a board in its test position, although complexity of wiring and even the simple physical fact of probes being sufficiently large to handle, each create their own limitations. As a consequence, fixtures are often available only up to specific sizes. Generally, the closer the test point probe centres, the smaller the fixture which is available.

Such a limitation may be no problem where boards of only low- to medium-densities of components are to be tested. Fixtures can, after all, even with pressure and size limits, have many thousands of test point probes. However, where boards with high-densities of components are to be tested, particularly if these components are very small (say, surface mounted very large scale integrated circuits, and so on), bed-of-nails fixtures may not allow the numbers of probes a circuit requires for adequate test. Fortunately, this limit is not reached very often.

Cost of fixturing is an important factor regarding bed-of-nails fixtures. They are quite expensive to produce: both to make and to program. Consequently, they are typically used by manufacturers of printed circuit boards made in medium- to high-volumes, where cost may be written off over large numbers of products. They do, however, provide a satisfactory fixturing method where large numbers of individual tests are to be performed over large numbers of circuit access nodes, and rapid testing is required.

### Dual-level fixturing

Occasionally a single bed-of-nails fixture is used to provide two levels of test, by using test point probes with different spring travel lengths. Thus, a board fully actuated into its test position has all test point probes located onto all required circuit node test points. Then, after allowing the board to move away from the test point probes with a short spring travel length, only those test point probes with a long spring travel length remain located onto circuit node test points.

### Bare-board testing

Bed-of-nails fixtures for bare-board testing are similar to those for assembly board testing, except bare-board testing requires a test probe to be positioned at each feedthrough hole on the circuit board. Consequently, numbers of required test point probes are significantly higher – around three to ten times those required for an assembled board fixture.

### Dual-fixture testing

In test situations where time is of prime importance, time taken by an operator to install a printed circuit into a bed-of-nails fixture, then remove it after test, may be eliminated by using two identical fixtures on one automatic test equipment system. Such test station setups are known as **dual-fixture testers**, or **dual-chamber fixtures**. Then, as one board is being tested on one fixture, another is being removed, then a third installed on the other fixture. After tests on the first board are completed, the automatic test equipment system switches to the other fixture and tests the third board while the operator removes the first board and installs a fourth.

## Moving probe fixtures

Cost of fixturing using bed-of-nails fixtures is quite high. Apart from costs involved in making the fixture itself, automatic test equipment systems which use them must be programmed accordingly – which again costs money. One way round this is to use moving probe fixtures, which effectively eliminate need of simultaneous access by perhaps thousands of test point probes onto a circuit board. Instead, two or more probes are moved around above the board, actually locating on the board only at required circuit nodes. Thus, no mechanical fixture is required – a board simply locates directly onto the automatic test equipment system, while no more than about four moving probes effectively take the place of literally thousands of stationary probes. Moving probe fixtures are common in test procedures for bare-boards and assembled products.

   This type of fixture is important in three ways. First, individual tests on components are undertaken serially – between each test, time is taken (perhaps up to about 0.5 s) for probes to move. to the next required positions. Compared with almost simultaneous bed-of-nails fixture tests this means total test time is considerably longer and depends almost entirely on the number of test steps. Second, accuracy of probe positioning is not limited to matrix centres like bed-of-nails fixtures – accuracy is as good as $\pm 50\,\mu$m. Consequently, printed circuit board layout design is not

**Photo 3.5** *Veritrak auto-probing fixtureless tester close-up, using moving probes to test unassembled printed circuit board (Bath Scientific)*

restricted to the typical 2.54 mm (0.1 inch) matrix required by bed-of-nails fixtures. Finally, programming is inevitably much simpler and no complex mechanical fixture is required, meaning total cost of a fixture is much lower.

These features make moving probe fixtures attractive to manufacturers of printed circuit boards made in low volumes, particularly where a number of different products are manufactured. In such cases, high costs associated with bed-of-nails fixtures (making the fixtures, programming the tester) may be prohibitive, while the quite long total test time of the moving probe fixture is no particular disadvantage.

**Photo 3.6**   *Close-up of APT-2200N fixtureless tester, using moving probes to test assembled printed circuit board (Contax)*

## Electrical considerations

Fixtures, being electromechanical in all but a few instances, are often given second place when looking at the technicalities of automatic test equipment. However, there is much to be considered. They form an integral part of any test system and can affect results obtained significantly.

Automatic test equipment operation speed is sufficiently fast that any degradation of signal quality either from tester to tested product or from tested product to tester is likely to cause at best – imprecise, at worst – wrong, results to be obtained. Effectively, an automatic test equipment system cannot do the job it is intended to do if its fixture is not able to support its capabilities.

Looking at a bed-of-nails fixture as a worst case example, we can see that a problem arises because the path taken by any particular signal between tester and fixture, along a wire and through a test point probe onto the tested board then similarly back again, is of a significant length. Also, many other paths formed by similar wires and test point probes are quite closely situated. Obviously, signal paths will exhibit capacitance, inductance and resistance. There may be crosstalk between closely situated wires, and wires will certainly exhibit a limited bandwidth. Effects vary according to type of signals used in the test procedure, too. A small signal, such as a digital data stream or audio waveform, may suffer from crosstalk from another signal or a loss of high frequency content. On the other hand, a large signal, such as a power supply voltage or a high energy pulse (used in a process to electrically isolate parts of the tested circuit – see Chapter 5), may cause crosstalk to a closely situated small signal path.

With such effects in mind, we can isolate three specific examples where fixtures may affect signal quality:

- When a test system applies a small signal waveform to a tested board.
- When a tested board produces a result in the form of a small signal waveform, to be measured by a test system.
- When a test system applies a large steady signal or pulse to the tested board.

## Mechanical considerations

There are significant mechanical considerations to consider for bed-of-nails and moving probe fixtures. Most of these are concerned with physical properties of probes themselves, but printed circuit board design must not be forgotten.

### Probes

Typical probes used in electromechanical fixtures are known as **spring contact test probes**. A typical spring contact test probe is illustrated in Figure 3.2, comprising barrel, plunger and spring, and the barrel is crimped to prevent the plunger from coming out. It is spring-loaded such that pushing its plunger into the barrel compresses the spring, ensuring a corresponding opposing force which pushes the plunger back out again on release. Usually, probes are used in a fixture so they are compressed to their **working travel** when applied to a tested board – about two-thirds of total possible travel distance. Thus, allowance is made for mechanical tolerances in a fixture, which ensures both probes and tested boards are not damaged. Typical working travels of probes are between about 1.2 mm (0.05 inch) and 8 mm (0.01 inch), and are selected according to fixture application.

Metals used in probe parts vary. Typically, however, plungers are of heat-treated beryllium copper, turned to an accuracy of $\pm 5\,\mu$m (0.0002 inch). These are then electroless nickel plated, then gold plated. Occasionally, rhodium plating is used instead of gold, giving a harder finish; desirable for some applications.

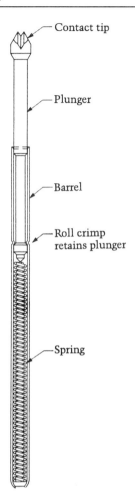

**Figure 3.2**   *Typical spring contact test probe, showing barrel, plunger and spring*

Barrels and receptacles are drawn from nickel silver, often pre-plated or alloyed with gold although phosphor bronze or beryllium copper are sometimes used where stronger parts are needed.

Spring material depends more closely on requirements: beryllium copper is used for lowest electrical resistance, stainless steel for high temperature applications, while music wire is used for highest spring force.

### Receptacles

Probes are used with **receptacles** which are permanently mounted into the fixture; a probe pushing simply into a receptacle, as shown in Figure 3.3.

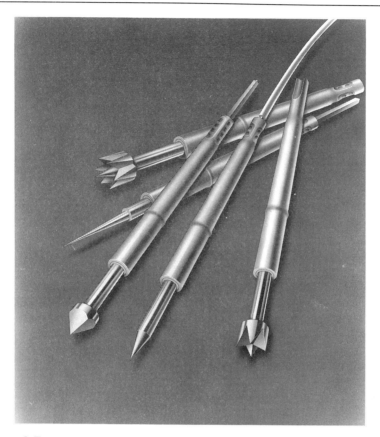

**Photo 3.7**   *A selection of probes used in bed-of-nails fixtures (Coda Systems)*

Receptacles are permanently positioned and wired into a fixture, and various wiring termination styles (push-fit, solder, crimp, wire-wrap and so on) are used to suit specific applications. Such a push-in arrangement allows probes to be changed without the trouble of rewiring or disturbing a fixture. One or more **detent crimps** in the receptacle side hold the probe into position while a fixture is in use and allow electrical contact between probe and receptacle.

Receptacles are generally held in a fixture by press-rings in the receptacle wall, which collapse as receptacles are pushed into a fixture hole. This fitting method requires a fixture plate of sufficient thickness. Alternatively, where fixture plates are too thin, fixture holes may be slightly larger, and press rings are then used as a simple shoulder to prevent a receptacle falling through the fixture. In such cases, however, receptacles must be glued in position.

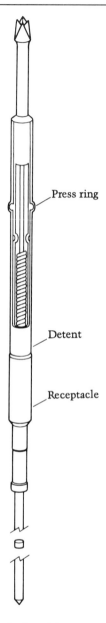

**Figure 3.3** *Probes are push-fitted into receptables. Receptacles are permanently fitted into fixture plates*

## Spring rate

One of the most important variables in determining which particular probe to use in an application is a probe's **spring rate**. This is simply the force presented by the spring against compression of the plunger into the barrel. Usually it is measured at two points in the plunger's travel: **initial spring rate** is the force required to move the plunger as compression first starts; **working travel spring rate** is the force required to hold the plunger at its working travel.

Spring rate is important in two ways. First, the greater the spring rate, the more effective is the probe in penetrating possible contaminants on the surface of the test board. Second, the greater the individual probe spring rate, the greater the overall pressure required to test a circuit board (working spring rate of one probe, multiplied by the total number of probes in a fixture). Obviously, a compromise must be reached between excessive spring rates and fixture actuation methods.

## Probe tips

Another important variable when selecting probes is the tip shape. A number of shapes exists, each having advantages in different applications. Several types of circuit nodes on a printed circuit board may exist, such as solder pads, plated-through holes, surface mounted components, conformally coated pads and so on. It is important to use the correct probe shape for each type, to ensure good electrical contact is made between probe and node. Figure 3.4 shows main probe tip shapes (although many others exist), while Table 3.1 lists those shapes together with typical applications.

## Surface mount assemblies

Surface mount assemblies create a specific problem when considering fixturing by probe means. Such assemblies usually feature high densities of components, and each component is extremely small. Consequently, probes themselves must be extremely small, and may have to be on a smaller matrix grid than other applications.

Fixtures for surface mount assembly test often, therefore, feature probes of a slightly different design than those discussed so far. Two methods are used: either mounting the probe plunger directly inside the receptacle, allowing fixture matrix grids down to about 1 mm centres (0.04 inch); or not using a receptacle at all, allowing fixture grids down to about 0.5 mm centres (0.02 inch). The first method allows plungers and hence tip shapes to be changed as required; while the second, although allowing even smaller sizes of probe to be made, fixes barrels and plunger directly into fixtures – so preventing tip change.

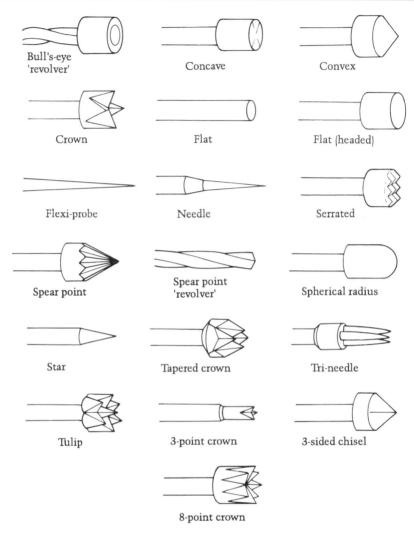

**Figure 3.4** *Typical probe tip shapes. Applications of probe tips are listed in Table 3.1*

## Coaxial probes

A special type of probe formed on a coaxial basis, with separate inner and outer probes, may be used for suppression of potential noise where high-frequency analog or high–speed digital circuits are tested. Coaxial probes are also useful in guarding methods used in electrical isolation test procedures (see Chapter 5), where two probe connections are often required at particular circuit nodes.

**Table 3.1** Typical spring contact test probe tip shapes, together with applications

| Tip shape | Application and comments |
| --- | --- |
| Bull's-eye revolver | dirty circuit nodes |
| Concave | long leads, terminals, wire-wrap posts |
| Convex | plated-through holes – leaves no indentations or marks |
| Crown | pads, leads – good penetration, self-cleaning |
| Flat | gold edge fingers – leaves no indentations or marks |
| Flat headed | gold edge fingers – leaves no indentations or marks |
| Flexi-probe | all nodes – good penetration of dirt and coatings |
| Needle | pads – good penetration of dirt and coatings |
| Serrated | pads, leads, terminals – periodic tip cleaning required |
| Spear point | pads, plated-through holes |
| Spear point revolver | dirty pads |
| Spherical radius | gold edge fingers – leaves no indentations or marks |
| Star | plated-through holes, pads – self-cleaning |
| Tapered crown | pads, leads, holes – self-cleaning |
| Tri-needle | pads, leads – good penetration of dirt, self-cleaning |
| Tulip | leads, wire-wrap terminals, pads – self-cleaning |
| 3-point crown | pads, leads, particularly surface mount components – self-cleaning |
| 3- or 4-sided chisel | plated-through holes – good penetration of dirt |
| 8-point crown | pads, leads – good contact reliability |

### Switch probes

Occasionally in a test setup, automatic test equipment needs to know if a particular part is present or absent. Switch probes contain a terminal, isolated from the remainder of the probe until the plunger operates and touches it. Thus, where a part is present the plunger hits the part, moves up the barrel and contact is made. On the other hand, an absent part causes no movement on plunger and contact is not made.

### Rotating plunger

By pressing a helix guide slot into a probe barrel, and using a plunger with a locating pin, probe plungers can be made to rotate as probes are compressed. This feature is useful where it is known circuit nodes may be dirty. As the plunger rotates it cuts through the dirt to ensure good contact on the node underneath. Typically, plunger rotations of 90° between initial and working travel are used in such applications.

# 4  Test strategies

Before considering processes used in automatic test equipment systems, it is essential to relate them – however roughly – to tests undertaken. Is the test a purely functional test, telling the user if a product works or does not? Is a full test of every component part essential? Are these component part tests carried out prior to assembly, or after assembly? What is expected of automatic test equipment – do you want it to simply tell you a product works or not; or do you want to know which component is faulty? Is speed of test more important than accurate fault diagnosis?

Determining such questions and answering them is not easy. They all form a basis for what should be any manufacturing company's test strategy. And, as all companies have different ideas of what automatic test equipment systems should do, there are consequently many different test strategies which may be followed. Indeed, there are probably as many different test strategies as there are manufacturing companies because each manufacturer's requirements are unique. Fortunately, these different test strategies can be roughly grouped into just a few basic categories.

A study of available and future test strategies is essential, in order that a correct and decisive choice may be made regarding automatic test equipment systems. Without considering test strategy on a long-term basis, manufacturers may find themselves with obsolete yet expensive test equipment, which cannot provide adequate test coverage of products manufactured at a later time.

A single solution for any product is rare; usually a range of options occurs, so manufacturers must be able to understand what these options provide. Only then may automatic test equipment be costed.

In a vast majority of tests, products comprise assembled and soldered printed circuit boards, the essence of which is a substrate into or onto which possibly hundreds of electronic components are soldered. So, in one simple step we have turned a single product into literally hundreds of individual items to be tested. There are three main ways which products may then be tested: functional test, in-circuit test, combinational test. Descriptions of automatic test equipment types which follow these three main ways are found, together with their derivatives, in Chapter 2.

## Functional test

A single **functional test** (illustrated in Figure 4.1) tells us if the product is working or not – and if it does work, all well and good. But if a product doesn't work, a functional test cannot easily tell us which component (or components) isn't working: in other words, a basic functional test can't tell us how to repair the product. Often, functional test is said to have **poor diagnostic resolution**, a factor which may not be important in late production stages, say, where a television receiver has been assembled and a final functional test is required just to prove it displays a picture – at which time, after all, manufacturers aren't looking to locate individual faulty components. However, level of diagnostic resolution *is* important earlier on in production, say, on assembly of each circuit board within the television receiver – then manufacturers need to determine and isolate faulty components, allowing repair. This fact may preclude use of functional test as a strategy.

During functional test an automatic test equipment system emulates a tested circuit's electrical environment, applying typical input signals then observing and evaluating output signals. Usually, a functional tester undertakes a test in two distinct parts:

- Verifying whether tested circuit works or not – often known as a **go/nogo test**.
- Circuits found not to work are checked to see if faults can be determined – typically operators follow step-by-step instructions for manual placement of measurement probes. Alternatives to manual

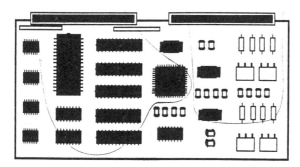

**Figure 4.1** *Showing basis of functional test, in which a product is tested overall*

**Photo 4.1**  *Electro-optical functional test system (Racal)*

probe placement are available, in the forms of movable guided probes or bed-of-nails fixtures. Ability to perform fault determination to any great diagnostic resolution depends largely on complexity of the automatic test equipment system; however, some functional testers do not feature this ability at all.

Fairly obviously a functional automatic test equipment system is structured exactly to match its tested circuit. It is circuit-specific, in that it cannot be used directly to test circuits of other origins without extensive adaptation. Adaptation from one tested circuit-type to another requires different fixturing between circuit and test system, as well as new controlling software.

   In matching a functional tester with a tested circuit the circuit's total function must be clearly and unequivocably understood by the tester's programmer. This relies on detailed information exchange between circuit designers and test equipment programmers. An important point here, is that programming time increases rapidly according to level of diagnostic resolution required (see later) – and time costs money. So, although a functional tester may be able to resolve specific faults, it may not be economical to use it, simply because of programming cost.

## In-circuit test

Pure functional testing looks at all circuit functions in a single process (the go/nogo test): determination of faults if they occur is usually then a matter of manual involvement by an operator. Automatically isolating those parts

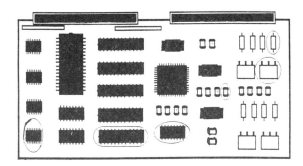

**Figure 4.2** *Showing basis of in-circuit test, in which individual components in a product are tested*

of the circuit which don't work (hence indicating how to repair faulty products), on the other hand, is the job of **in-circuit testing** processes (illustrated in Figure 4.2). Naturally, to do this, in-circuit testing has to look at each part of the circuit in isolation, generally in a serial manner; parts tested one at a time. In the end, though, in-circuit tests are simply testing whether individual components of a circuit are functional. Consequently, it is important not to confuse use of the term *functional* when applied to in-circuit testing.

At this point it is worth pointing out a sub-category of in-circuit testing known as **manufacturing defects analysis** (see Chapter 2), which

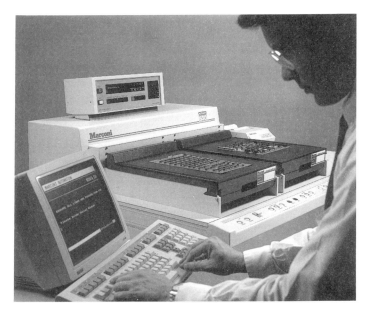

**Photo 4.2** *Midata 530 in-circuit tester (Marconi)*

**Photo 4.3**   *Midata 520 manufacturing defects analyser, shown with dual-chamber fixtures (Marconi)*

essentially is a means of locating manufacturing defects – those faults such as open or short circuits, or incorrectly mounted components, which occur as results of the manufacturing processes. The main fact which differentiates between manufacturing defects analysis and pure in-circuit testing is whether power is applied or not. In-circuit test applies power before determining whether devices are working or not; manufacturing defects analysis checks for faults without power applied. Although manufacturing defects analysis is technically a sub-category of in-circuit analysis, it is commonly considered an automatic test equipment type in its own right; automatic test equipment systems constructed around the type are known as **manufacturing defects analysers**. Because they are commonly found in a situation where all products are tested to eliminate manufacturing defects prior to testing with functional testers, manufacturing defects analysers are often known as **pre-screeners**, and the process of manufacturing defects analysis is known as **pre-screening**.

An advantage of in-circuit test over functional test is due to its inherent feature of isolation of circuit parts: total circuit function need not be understood by test equipment programmers. Programs are merely step-by-step tests of individual devices, all of which are probably held in a device test library. Further, adaptation of software from one type of tested circuit to another is quite simple and straightforward.

A further advantage arises from the step-by-step in-circuit test process as faulty devices are diagnosed immediately on their detection, so are pin-

**Photo 4.4** *DDS-40 fault diagnosis equipment for servicing or manufacturing defects analysis (ABI Electronics)*

pointed to the operator rapidly. It doesn't matter how many faulty devices are in a circuit, in-circuit test potentially detects them all in a single test run.

One disadvantage of in-circuit test, on the other hand, is an assumption that all individual devices, once tested and known to be working, are going to work together as a whole circuit. Indeed, in-circuit test relies on a tested circuit being 'perfect'. Of course, there is no corresponding assurance – all devices may work in isolation but a poorly designed circuit need not work in total.

Another disadvantage stems from the need for a complex fixture of the bed-of-nails type. Fixed test probes must exist on this fixture at every nodal access point required into the tested circuit. Thus an in-circuit fixture is usually difficult to make (a one-off construction) and is always specific to one type of tested circuit.

Fixtures depend on production technology used for a particular tested circuit assembly. One of the important factors has to do with test probe spacing, which is mechanically limited to spacings of no less than around 2.5 mm (0.1 inch). Chapter 3 describes this in detail. Where circuits are assembled using very large scale integrated circuits or, particularly, surface mounted components, this spacing may not be small enough to give adequate nodal access to ensure all components may be individually tested by an in-circuit tester. Functional test on the other hand, although

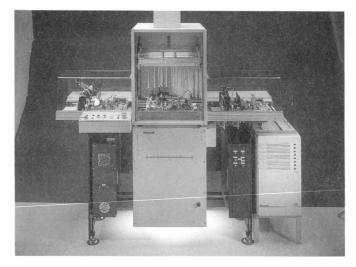

**Photo 4.5**   *Midata 520C compact manufacturing defects analyser (shown at the right of the setup), used for in-line testing (Marconi)*

overcoming nodal access problems, requires considerably greater involvement in terms of time, manpower and programming and may not give adequate diagnostic resolution. Consequently, a problem is created which neither in–circuit nor functional test alone are able to properly overcome.

## Combinational test

A common method of automatic test is to **partition** the circuit into smaller, more easily handled, parts (illustrated in Figure 4.3). These parts, commonly called **clusters**, may comprise any number of devices from, say, hundreds, to as few as one.

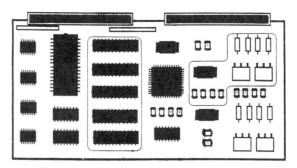

**Figure 4.3**   *Showing basis of combinational test, in which products may be partitioned and each partition tested separately*

**Photo 4.6** *S790 combinational test system (Schlumberger)*

Partitioning clusters from the rest of the tested circuit is a process of electrical isolation (see later for hardware methods of partitioning, and Chapter 5 for software methods), undertaken primarily by having adequate numbers of test points in the circuit. However, its aim is to test each cluster functionally. Cluster testing is therefore usually seen as a mixture of in-circuit testing (where each device is usually considered individually) and functional testing (where the circuit is usually tested as a whole). For this reason cluster testing is often known as **combinational testing**, and automatic test equipment systems performing it are called **combinational testers**.

Some combinational automatic test equipment systems feature ability to vary cluster size according to tested circuit requirements. Phrases coined for this ability are **polyfunctional testing**, **variable in-circuit partitioning** (VIP) or **multimode testing**. Variable in-circuit partitioning allows any partition size from single components, component clusters, complete sub-circuits through to entire assemblies to be tested according to requirements. This effectively allows a continuous capability from full in-circuit test to full functional test, chosen by the system programmer.

Such a test strategy is responsive to practical needs of the test environment. It allows test programs to be quickly developed, using typical in-circuit techniques, to test circuits to a device level. Later, high degrees of fault coverage, using functional test techniques, may be introduced to suit applications.

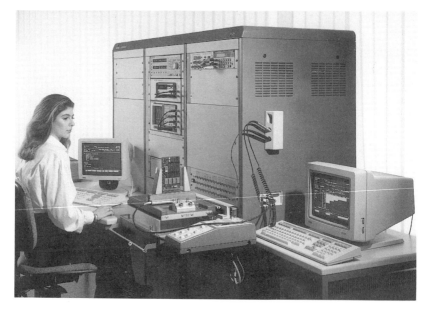

**Photo 4.7** *TSP combinational test system (Rohde & Schwarz)*

**Photo 4.8** *Midata 515 combinational test system (Marconi)*

**Photo 4.9** *Midata 560 combinational test system (Marconi)*

## Bench-top test

A relatively recent category of automatic test equipment is the **bench-top tester**. This is, however, extremely difficult to categorize in terms of test strategies, because there is no defined strategy which any particular bench-top tester follows. Rather, it depends on a manufacturer's idea of what a bench-top tester should be, usually perceived as a market requirement not a definite test strategy. Consequently bench-top testers range in complexity from equipment which tests individual components, through in-circuit test, functional test, to combinational test.

## Design for testability

Basic test methods: functional test, in-circuit test, and combinational test are not, alone, able to solve all test problems if products are complex and miniaturized circuits where access to all circuit nodes is impossible. Very large scale integrated products and, specifically, surface mounted assemblies which both may feature extremely dense circuits do not always allow total nodal access with test point probes. This trend towards miniaturization using such assembly methods is accelerating, too; a fact which will cause more problems as time progresses. Already it has been

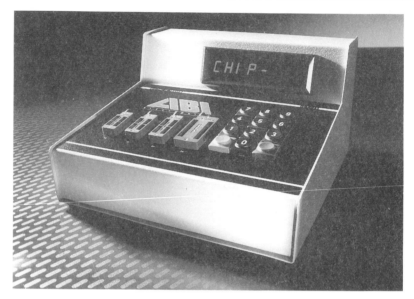

**Photo 4.10** *ICT-24 digital IC tester (ABI Electronics)*

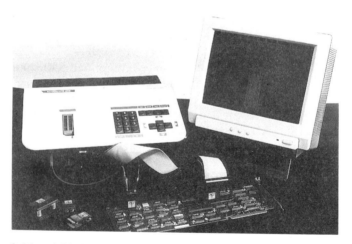

**Photo 4.11** *ABI Junior digital IC in-circuit tester (ABI Electronics)*

**Photo 4.12** *SI635 diagnostic board tester (Schlumberger)*

**Photo 4.13** *Midata 511 bench-top combinational test system (Marconi)*

**Photo 4.14**   *TSA bench-top test system, providing in-circuit, functional and combinational test facilities (Rohde & Schwarz)*

determined that, while in–circuit testing of small to medium scale integrated products detects nearly all possible faults, only around 85% of faults in large scale integrated products, and somewhat less than 75% of very large scale integrated product faults, may be detected by in–circuit means.

While a functional test strategy may be able to detect more faults than this, programming requirements and test times both rapidly become excessive as products increase in circuit density and complexity, with a consequent and dramatic cost difference.

With this in mind manufacturers and designers are turning their attention, on a wider scale, to building certain features in to their products to allow following test stages to be undertaken more easily and, even more important, with greater effect and more cheaply. In the main, these features are intended to allow partitioning of tested circuits into smaller functional blocks; with a similar effect of test ease seen earlier in combinational test strategy. However, partitioning in combinational test is done under automatic test equipment system software control; features considered here are hardware additions.

Generally, inclusion of such features is called **design for testability** (DFT) which, by the fact of its virtue in aiding product testing, is a form of test strategy in its own right. Two main methods are identifiable: *ad hoc* measures and structured approaches.

## Ad hoc *design for testability*

*Ad hoc* measures are those methods incorporated by a circuit designer, when designing a new product, to enable easier testing after production. Usually, therefore, they are fairly simple measures, often undertaken as general guidelines – whenever a particular device is used, the designer incorporates corresponding measures to suit the device.

Often *ad hoc* measures may be summarized in a collection of rules which the designer must follow, for example:

- Ensure adequate test point access in circuits. This is fairly self-explanatory; if insufficient test point access is available in a circuit, automatic test equipment cannot isolate individual circuit devices or parts.
- Where device connections are unused during normal use, ensure any which may show internal device operation are available to a test point. Key signals within very large scale integrated circuits are often available at device connections which, though they are unused in normal operation, may be extremely useful to automatic test equipment to aid in fault detection.
- Ensure memory elements' reset and preset lines are accessible through test points. Although devices within a particular circuit may not require use of reset or preset lines in normal operation, testing is generally made much easier if devices can be reset or preset with a single pulsed input. If this facility is not available, test equipment usually has to generate an initialization sequence to ensure devices are in a known state (see Chapter 5).
- Use resistors or logic gate output to tie devices to fixed potentials. Connections for external instruction access, initialization lines (such as reset and preset lines – see earlier), and interrupt lines are all examples of device connections which, under normal operating conditions, are often unused and so are tied to fixed potentials. However, rather than hard-wiring these connection lines directly to fixed potentials (as shown in the counter in Figure 4.4a), resistors or logic gate outputs should be used to do this (as in the more testable counter in Figure 4.4b). As before, simple pulses can then be used to initialize tested circuits, rather than complicated initialization sequences.
- Ensure means of isolating feedback loops are incorporated. Where feedback loops exist as shown in the example circuit in Figure 4.5a, even if adequate test points and initialization lines are accessible to the automatic test equipment system, feedback loops mean any fault may still be propagated around the loop. To overcome this situation all loops should be breakable on instruction from the test equipment, as suggested in the circuit in Figure 4.5b where a gate is added.
- Where access to reset and preset lines or feedback loop isolation is not

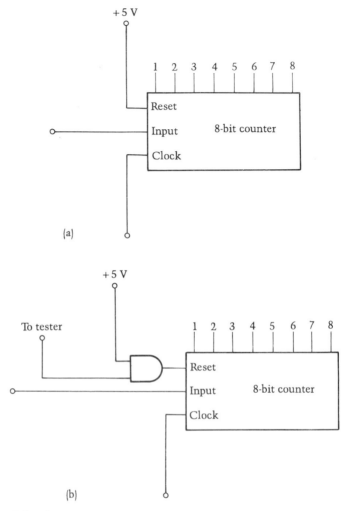

**Figure 4.4** *Incorrect and correct design (a) a device with tied instruction lines, initialization lines or interrupt lines hard-wired to fixed potential (b) tying to fixed potential via resistors*

possible (say, in very large scale integrated circuit devices) ensure initialization sequences may be applied. If all else fails, and additional initialization pulse input test points are not possible, ensure automatic test equipment initialization sequences may be applied to all devices.

- Allow facilities enabling electrical partitioning of a tested circuit into a number of sub-circuits. It is often possible to add simple gates to a circuit to allow simple partitioning of a tested circuit, which can improve testability and decrease test times considerably. A 16-bit

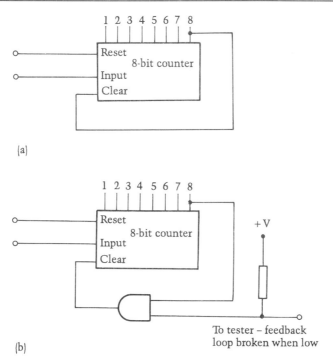

**Figure 4.5** *Incorrect and correct design (a) a feedback loop (b) simple method of breaking a feedback loop*

counter is shown in Figure 4.6a. To initialize the counter to a state where all bits are logic 1 requires a reset-to-0 pulse, followed by a count of 65,536 clock pulses – following devices driven by the counter thus require this count for each test stage. Obviously later devices will take a considerable test time. By adding gates in between counter stages, as shown in Figure 4.6b, the counter is effectively broken up into two 8-bit counters, which each may be tested in just 256 test clock cycles. Adding more gates after the counter allows following devices to be tested in total isolation from the counter, too. Electrical isolation of circuit parts in this manner, although requiring extra gates and making the circuit marginally more complex, is merely a matter of added hardware and is significantly easier and cheaper than other methods of electrical isolation, which rely on automatic test equipment system facilities (described in Chapter 5).

These are fairly basic measures which give an idea of the concept of design for testability. Once such *ad hoc* rules are standardized within an organization, automatic test equipment system programming and use becomes more standardized, too.

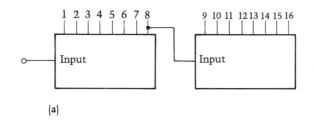

(a)

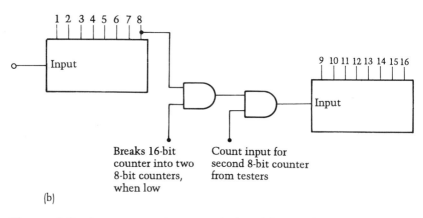

Breaks 16-bit
counter into two
8-bit counters,
when low

Count input for
second 8-bit counter
from testers

(b)

**Figure 4.6** *Incorrect and correct design (a) complete counter (b) breakable counter*

### Structured approaches to design for testability

While *ad hoc* measures can considerably increase a particular circuit's testability, they are specific to a circuit and may not necessarily be usable in other circuits. Structured approaches to design for testability, on the other hand, are generally applicable to all circuits. They rely, however, on manufacturing-stage additions to special devices used in a circuit, rather than to design–stage additions incorporated into circuits with ordinary devices. Two main methods of structured design for testability are identifiable: scan test and built–in self–test, although many types of each exist.

### Scan test

The basis of scan test is addition of the ability of an integrated circuit to be switched into a test mode (with a test/normal input pin connection) in which internal input and output latches are all connected together to form a shift register, known as the **scan register**. **Scan in** input and **scan out** output pin connections, directly to and from the register, are also provided, and scan tests on individual integrated circuits are often called **internal scan tests**.

Scan testing an integrated circuit with a scan register can merely entail scanning a test pattern vector into the register, returning to normal operation for a defined number of cycles – usually one – then re-establishing test mode and scanning out the resultant vector from the register. Other test pattern vectors may be scanned in and out to test individual functions of the integrated circuit, but the overall process is as simple as this single test stage shows.

Scan input and output pin connections of integrated circuits within a whole product may be linked together, forming a complete scan register around the circuit. Because of this, a circuit design using scan test devices is also known as a **boundary scan design**. With boundary scan design a complete circuit's test process can, in effect, be as simple as scanning in a single input vector and observing the output vector. The vector output pattern immediately allows determination of faults, isolating particular integrated circuits where faults have occurred. In practice, many vectors need to be scanned in and out if sequential logic elements are in circuit.

Scan test design is useful because its very presence effectively breaks up a circuit into device-sized partitions during test. Direct physical access to devices, with nodal test point probes for in-circuit test strategies, is not required. This means device-level testing of miniaturized products, in the form of very large scale integrated circuit and surface mounted component assemblies, is immediately catered for and easily undertaken if all devices in assemblies are scan test devices.

Structural testing of a circuit assembly: testing interconnections between devices (those manufacturing defects – such as opens or shorts – which form a large proportion of all assembly faults) is made much easier with a scan test strategy. The basis of this is: as a scan vector is shifted out of a scan register it is immediately apparent, due to a break in the vector, where an interconnection fault is situated in the circuit. Many variations of this are currently being developed.

Initialization of a circuit to any desired state is comparatively simple, too, as test pattern vectors can be chosen to do this in one short stage. Long initialization patterns, propagated step-by-step through a complex sequential circuit, are not needed.

Where circuits comprise scan test devices as well as conventional devices and components such as resistors and capacitors, scan testing can provide a means whereby these non-scan test components may be tested in a single test process.

Scan test strategies are currently defined in the American IEEE provisional standard P1149.1, having been developed by an organization called Joint Test Action Group (JTAG) comprising a number of integrated circuit manufacturers.

This defines a scan test integrated circuit as shown in Figure 4.7. Here, four connections to the integrated circuit are common, and so form what is known as a **four-wire testability bus**. Two control signals: **test clock** (TCK); **test mode select** (TMS), together with scan register in and out connections; **test data in** (TDI); **test data out** (TDO), make up the four-wire testability bus. A simple connection of one integrated circuit's TDO connection to the TDI connection of the next integrated circuit allows test information to be scanned through both. Functional circuits within any integrated circuit are essentially unaffected by the four-wire testability bus, except for the inclusion of input and output latches at each remaining connection pin.

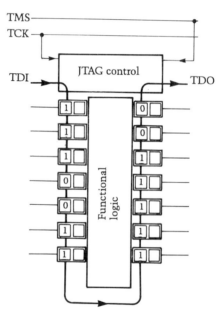

**Figure 4.7**   *Scan test integrated circuit showing four-wire testability bus*

Interestingly, IEEE P1149.1 only specifies general mechanical and operational features. What is scanned through an integrated circuit as a test pattern vector is undefined, and so up to the system designer. Data passing through a scan register may be straightforward test data, initialization data, or potentially most important, test instructions.

Scanning of test instructions means that integrated circuits with internal test features may be controlled as part of a scan test. Thus, automatic test equipment systems are able to control such features simply – without need of elaborate bed-of-nails fixtures or initialization sequences.

In general, scan testing appears to offer solutions to many problems currently encountered in automatic test equipment systems. In particular, programming may be more easily undertaken, test times may be reduced and expensive fixturing is eliminated. Dense, miniaturized circuits may be tested for structural faults to a level which in-circuit testing cannot reach. This can make the functional tester's task much simpler – merely requiring a go/nogo test to determine correct operation.

### Built-in self-test

Scan test techniques can reduce requirements of complex test patterns generated by automatic test equipment, but does not eliminate them. For this reason, integrated circuit designers are currently proposing a number of additions of internal tests for integrated circuits. Such tests are called **built-in self-tests** (BISTs) and many schemes are already available, following many methods.

One of the most common built-in self-test methods, incorporated into several schemes, uses two **linear feedback shift registers** (LFSRs) in addition to main functional circuits within an integrated circuit. The first linear feedback shift register generates a sequence of test pattern vectors which is applied to the main functional circuits, while the second generates a signature from the resultant output. Basic signature analysis procedures within the automatic test equipment compare these signatures with expected results to verify correct operation or detect if an integrated circuit is faulty.

In such integrated circuits with built-in self-test, these linear feedback shift registers will normally double as scan registers for use in scan test. Coupled with scan test, signatures are naturally routed as test pattern vectors directly to the automatic test equipment.

### Mixed test requirements

From the automatic test equipment point of view, most test requirements for the foreseeable future are going to have to deal with tested circuits comprising integrated circuit devices of scan test, conventional and, perhaps, built-in self-test designs. Coupled with these integrated circuits will be non-integrated active and passive components such as resistors, capacitors, coils, and semiconductors such as transistors and diodes. Although scan test strategies provide a means of simplifying test procedures for dense scan test design circuits, which would otherwise not be able to accommodate in-circuit bed-of-nails fixture probes, it is not immediately apparent how dense mixed boards of scan test, conventional, and built-in self-test designs will benefit.

Fortunately, however, techniques are currently being developed which use scan register output vectors to allow observation of non-scan test components.

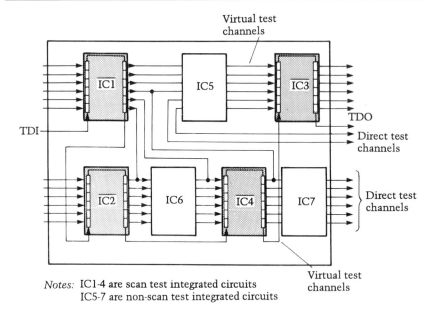

*Notes:*  IC1-4 are scan test integrated circuits
IC5-7 are non-scan test integrated circuits

**Figure 4.8**   *Printed circuit boards may contain scan test and non-scan test components. At the junctions between, virtual test channels are available*

Figure 4.8 shows a potential printed circuit board arrangement, with scan test integrated circuits alongside non–scan test components. The very fact that an interface occurs between the two component types means access is possible to signals around the non–scan test components. In effect, signals at each interface between scan test components and non–scan test components may be held by scan registers on the scan test components, and scanned out to the automatic test equipment system. Thus clusters of non-scan test components are isolated with these **virtual test channels**, and may be tested by standard means.

## Test area management systems

While strategies for automatic test equipment, and tests it performs are numerous and reasonably well documented, one strategy currently receiving not so much attention is its management. Significant amounts of money are involved in purchase, operation and maintenance of automatic test equipment, yet all too often the equipment is used in a stand-alone situation, barely scratching the surface of the ability it can have in a total test area system.

Granted, data generated by an automatic test equipment system is mainly useful in detection of faulty products. That is, after all, what it is bought for in the first place. However, it can have other, associated uses. Some of these are:

- Statistics gathering and analysis – information relating to faults; where they occur, how regularly they occur, what components are involved, identification of faulty circuits and so on is all highly relevant, particularly in medium- to large-volume production runs. Even where small-volume product quantities are involved, it is useful to appreciate test area trends on an objective statistical basis rather than as a subjective assumption.
- Computer-aided repair – after test, boards found faulty must be repaired. Information from automatic test equipment is inevitably helpful in the following repair stages. Systems incorporating computer-aided repair correlate this information with the faulty product, allowing repair personnel access as required.
- Real-time warnings – commonly known as **watchdogs**, real-time warnings may be used in systems to indicate, with visual or audible means, specific preset production parameters.
- In a manufacturing database – large systems incorporating databases may be matched with, and allow access by other related computer-aided parts of the design, development and production cycles. Thus, the whole manufacturing phase can be integrated with the aims of coordination and streamlining.

Where systems incorporate such facilities, they are commonly called **test area management systems** and essentially form a basis for computer-aided engineering systems.

## A total company test strategy

Whatever a company makes, it must test. The extent to which products are tested depends on company priorities. If the company is out to make a fast buck, testing may be little more than a quick visual check. If customer satisfaction and product reliability is the aim, testing is usually thorough, often time-consuming and always expensive (taken in isolation, that is).

Consequently, prior to evolving a total test strategy along with its resultant test requirements, each company must decide its priorities. These are, naturally, outside the scope of this book and, at first sight, may lead you to assume that it is impossible to discuss total test strategy at all. However, guidelines can still be given which are, when all's said and done, just basic pointers to the way a company chooses to go.

Generally, there are four main factors which must be initially considered:

- Percentage of circuits expected to work, without need of repair. This is highly dependent on quality of assembly – the higher the quality, the higher the percentage expected to work correctly. Often the percentage figure is given as the **first-pass yield**, relating to the numbers of products which work on their first pass through automatic test equipment. High first-pass yields are those over 80%, medium yields are between 40 and 80%, while low yields are below 40%.
- Types of faults expected. Faults are categorized into a number of groups such as manufacturing defects, device operation faults, performance faults.
- Production volume. Numbers of circuit assemblies made by a company in a year. These are usually grouped into fairly loose amounts as high-volume (over 500,000), medium-volume (100,000 to 500,000) and low-volume (less than 100,000).
- How many different circuits are to be produced. Companies manufacturing circuits in high-volume production are, typically, unlikely to be producing more than just a handful (say, less than 10) of different types of products. Mid-range circuit numbers of around 40 circuit types are common, while at the other end of the spectrum, manufacturers of circuits in low-volume production runs may produce many tens (say, a hundred) of different product types.

These factors interact considerably, each having a bearing on the type of automatic test equipment used. Even these loose categories of each factor give large numbers of possibilities. Taking these examples (three categories for each factor) gives some 81 permutations ($3 \times 3 \times 3 \times 3$) of possibilities, which must all be catered for as a company chooses its automatic test equipment system. If a company wishes to define more categories within each factor, a corresponding greater number of permutations evolves. However, this is probably unnecessary and 81 will suffice, at least for our purposes here!

## Combining automatic test equipment systems

In many production environments it is not sufficient to have just one item of automatic test equipment. Where products are particularly complex, or where volumes are reasonably high it is often essential to combine equipment to maintain steady and accurate testing procedures.

Equipment is combined either in serial or parallel, making use of the inherent features of whatever equipment is used. Thus circuits may be rapidly tested using a manufacturing defects analyser, and only those boards which are found to be faulty are passed to an in-circuit tester for fault identification (a serial test strategy). Or, circuits may be tested by one of a number of functional testers (a parallel strategy). However, it is not unusual to see multiple combinations of serial *and* parallel automatic test equipment, depending on company requirements.

## Serial combinations

Where two or more testers are combined in serial, the first tester is commonly known as a **screen** tester, as it screens products from following testers. In other words, the first tester merely checks for common faults which can be tested simply. If products with such faults are allowed to pass to a second and subsequent testers without screening, unnecessary time is wasted by later test stages identifying simple faults.

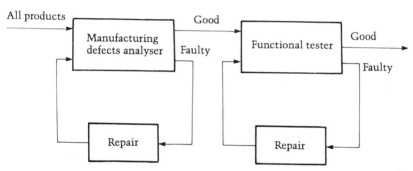

**Figure 4.9** *Screening, more commonly called pre-screening, by a manufacturing defects analyser in serial with a functional tester*

Screening or, as its is often known, **pre-screening** is illustrated in Figure 4.9, where a manufacturing defects analyser is shown in serial with a functional tester. All products are tested by the manufacturing defects analyser, so any found to have a manufacturing defect can be repaired prior to testing by the functional tester. Thus the functional tester is allowed to get on with more complicated testing procedures and is not required to detect simple faults.

## Parallel combinations

Figure 4.10 shows a test situation in which all products manufactured are tested by two functional testers in parallel. This type of combination may be required if product volumes are sufficiently high that a single tester has a test time greater than a product's assembly time. A parallel combination of two testers effectively halves overall test time.

Which combination of automatic test equipment is chosen, indeed whether a combination is necessary at all, depends totally on the four

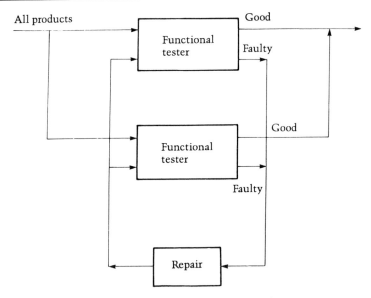

All products

Good

Functional
tester

Faulty

Good

Functional
tester

Faulty

Repair

**Figure 4.10** *Parallel testing with two functional testers*

factors noted earlier. Table 4.1 is an array of all 81 permutations of the four factors influencing automatic test equipment which may be used, either individually or in combination, are given. These are, however, merely examples and although fairly comprehensive within the limits of such a table's production, are not meant to be exhaustive. The table can be, on the other hand, a starting point and offers some idea of how to evolve a total test strategy. Final choice of a test strategy is, though, dependent on an individual company's requirements and how permutations of factors and indeed, how those factors themselves, are viewed by the company.

Table 4.1 lists automatic test equipment systems suggested in any given permutation of factors as weighted terms. These weighted terms do not indicate actual numbers of individual items of automatic test equipment – this would be impossible to define, as each item of automatic test equipment made by different manufacturers within any category has different facilities and features. Instead, terms merely imply an indication of importance of those items of automatic test equipment. Equipment is grouped into four main categories: $M$ meaning manufacturing defects analysers; $I$ meaning in-circuit testers; $F$ meaning functional testers; $C$ meaning combinational testers. Importance is defined in an ascending range from 1 to 5, where 1 is least important, 5 is most important.

So, as an illustration, a term $5M + 4C$ indicates tests undertaken by manufacturing defects analysers and combinational testers are extremely

**Table 4.1** Automatic test equipment system types, as weighted terms in an overall test strategy according to the four main fault factors

*Circuit mix (numbers of different circuits)*

| | High first-pass yield | | | Medium first-pass yield | | | Low first-pass yield | | | |
|---|---|---|---|---|---|---|---|---|---|---|
| | 100+ | 40+ | 10 | 100+ | 40+ | 10 | 100+ | 40+ | 10 | |
| **High-volume** | 5M+3I+4F / 5M+4C | 4M+2I+3F / 4M+3C | 3M+1I+2F / 3M+2C | 4M+3I+4F / 4M+4C | 3M+2I+3F / 3M+3C | 2M+1I+2F / 2M+2C | 5m+3I+3F / 5M+5C | 4M+2I+2F / 4M+4C | 3M+1I+1F / 3M+4C | Manufacturing faults |
| | 4M+5I+4F / 4M+5C | 3M+5I+3F / 3M+4C | 2M+4I+2F / 2M+3C | 3M+5I+4F / 3M+5C | 2M+5I+3F / 2M+4C | 1M+4I+2f / 1M+3C | 4M+5I+3F / 4M+5C | 3M+4I+2F / 3M+4C | 2M+3I+2F / 2M+4C | Device faults |
| | 3M+3I+5F / 3M+5C | 2M+3I+4F / 2M+4C | 1M+2I+3F / 1M+3C | 2M+3I+5F / 2M+5C | 1M+3I+4F / 1M+4C | 1M+2I+3F / 1M+3C | 3M+2I+5F / 3M+5C | 2M+2I+4F / 2M+4C | 1M+1I+3F / 1M+3C | Performance faults |
| **Medium-volume** | 4M+3I+3F / 4M+4C | 3M+2I+2F / 3M+3C | 2M+1I+1F / 2M+2C | 3M+2I+2F / 3M+3C | 2M+2I+2F / 2M+3C | 1M+1I+1F / 1M+2C | 4M+2I+2F / 4M+4C | 3M+2I+2F / 3M+3C | 2M+1I+1F / 2M+3C | Manufacturing faults |
| | 3M+4I+3F / 3M+4C | 2M+4I+2F / 2M+4C | 1M+4I+2F / 1M+4C | 2M+3I+3F / 2M+4C | 1M+3I+2F / 1M+4C | 1M+2I+2F / 1M+3C | 3M+3I+3F / 3M+3C | 2M+3I+2F / 2M+3C | 1M+2I+2F / 1M+3C | Device faults |
| | 2M+3I+4F / 2M+4C | 1M+3I+3F / 1M+4C | 1M+2I+3F / 1M+3C | 1M+2I+4F / 1M+4C | 1M+2I+3F / 1M+3C | 1M+1I+3F / 1M+3C | 2M+2I+4F / 2M+3C | 1M+2I+3F / 1M+3C | 1M+1I+3F / 1M+2C | Performance faults |
| **Low-volume** | 3M+2I+2F / 3M+4C | 2M+2I+1F / 2M+3C | 1M+1I+1F / 1M+2C | 2M+1I+2F / 2M+2C | 1M+2I+2F / 1M+2C | 1M+1I+1F / 1M+1C | 3M+1I+2F / 3M+1C | 2M+1I+1F / 2M+1C | 1M+1I+1F / 1M+1C | Manufacturing faults |
| | 2M+3I+2F / 2M+4C | 1M+3I+1F / 1M+3C | 1M+2I+1F / 1M+2C | 1M+2I+2F / 1M+2C | 1M+2I+1F / 1M+2C | 1M+2I+1F / 1M+2C | 2M+2I+1F / 2M+2C | 1M+2I+1F / 1M+2C | 1M+2I+1F / 1M+2C | Device faults |
| | 1M+2I+3F / 1M+3C | 1M+2I+3F / 1M+3C | 1M+1I+2F / 1M+3C | 1M+1I+3F / 1M+3C | 1M+1I+3F / 1M+3C | 1M+1I+2F / 1M+2C | 1M+1I+3F / 1M+2C | 1M+1I+3F / 1M+2C | 1M+1I+2F / 1M+2C | Performance faults |

M = manufacturing defects analyser
I = in-circuit tester
F = functional tester
C = combinational tester

important, although those undertaken using manufacturing defects analysers are marginally more important than those on combinational testers. Similarly, the term *3M + 2I + 5F* indicates functional tests are extremely important, while manufacturing defects analysis tests are quite important, and in-circuit tests are marginally important.

It is important to remember, when using Table 4.1, individual test strategies such as manufacturing defects analysis, in-circuit test, functional test and combinational test are not universally fixed or defined by all manufacturers, all of the time. One manufacturer's in-circuit tester may provide significant manufacturing defects analysis; another manufacturer's functional tester may give both in-circuit and manufacturing defects analysis support. In the end, automatic test equipment must be considered side-by-side, and comparisons made, so that any particular chosen test strategy can be followed to the best of the equipment's ability.

Using such a method of defining importance of tests and associated automatic test equipment, a basis for a suitable test strategy may be realized.

## Economics of overall test strategy

While the aim of an overall test strategy must be to produce a product of guaranteed performance, inevitably the level of this performance is a function of cost. In other words, how much money is it going to cost? There is no easy answer to this. Final cost of a test equipment system depends largely on what a company wants from it. Indeed, a question which may now arise is: do we really need automatic test equipment at all? Manual test equipment systems can, after all, perform test functions to an adequate if not perfect level.

Manual and automatic test equipment systems can be compared in terms of costs quite simply. First, we can consider purchase costs – capital costs. Manual test equipment systems are reasonably cheap to buy, while automatic test equipment systems are expensive in comparison. These costs are fixed and can be determined quite accurately. On the other hand, ongoing costs for both types of test equipment systems are variable and not so easy to determine. Ongoing costs for manual test equipment systems are, however, generally considered to be higher than those of automatic test equipment systems, although numbers of products being tested has an important bearing on the outcome of a cost analysis. Figure 4.11 shows costs of manual and automatic test equipment systems as graphs of total costs against product numbers being tested. While capital, fixed costs of purchase of automatic test equipment systems are higher than those of manual test equipment systems, variable, ongoing costs are generally lower. Although for low numbers of products being tested manual test equipment systems are cheaper than automatic test equipment systems, at

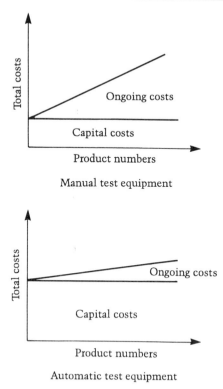

**Figure 4.11** *Comparing costs of manual and automatic test equipment systems*

some point, depending on the number of products tested, automatic test equipment becomes cheaper than manual.

Perhaps the best way of looking at it is to say that any test equipment should have the highest diagnostic capability for a particular fault type at the lowest recurring cost. Thus, test equipment chosen at any stage within the test strategy is the most economic for that position.

However, an *overall* test strategy should not be considered in isolation. Product testing is only one small part of the much larger topic of product quality. For any given product a company must have, consciously or unconsciously, a quality philosophy. At the two extremes a product has either a poor quality and is usually unreliable, or has a high quality and is usually reliable.

Customer satisfaction is a good measure of quality. If customers buy a product and find it useful, reliable, acceptably priced and finally, of acceptable quality, they are generally satisfied. From a marketing point of view, again either consciously or unconsciously, it is a manufacturer's task to determine these factors which relate to customer satisfaction.

## A scenario relating test strategy cost with quality

As an example of how product quality and cost of test strategy must not be viewed in isolation, the following scenario acts as a useful illustration.

Qwertyuiop Electronics plc, is an established manufacturer of television receivers, using manual test equipment processes in production. Current sales are around 20,000 sets a year, at an average price to the consumer of £500. Qwertyuiop televisions are sold to the public through retail outlets around the country, and 30% of the retail price is profit for the outlets themselves.

A warranty period of one year is given with all televisions, and approximately 10% of televisions sold require some service work in that time. Retail outlets service the receivers, and bill Qwertyuiop Electronics for work carried out under warranty. Average service charge to the company is £30 per repair.

Service expenditure:

20,000 receivers, 10% require service, average cost £30 = £60,000

is spent by the company, each year, to have television sets repaired.

The company installs automatic test equipment to the tune of £100,000 (which it intends to write off over a four year period) in an attempt to improve product reliability.

After the equipment is installed, an instant improvement is apparent, with only 3% of television sets requiring service work to be carried out during the warranty period. The service expenditure is now:

20,000 receivers, 3% require service, average cost £30 = £18,000

To demonstrate its improved product quality, Qwertyuiop Electronics decides to give customers a two year warranty period. During this time, a further 2% of receivers require service, so the service expenditure is now:

20,000 receivers, 5% require service, average cost £30
= £30,000 over two years
= £15,000 a year.

Thus, a saving of £45,000 a year has been made on service expenditure. The test and inspection equipment cost can be written off in just over two years – not the original period, thought, of four years.

However, as Qwertyuiop televisions now have a two year warranty period, a price increase of, say, 2% is justified. Extra income, therefore, is:

20,000 receivers, at 70% of 2% of £500 = £140,000

which means the company is well into profit at the end of the first year, simply because automatic test equipment was purchased.

Quality, goes the saying, is free. And, here, it gives a profit, too.

## Where to use automatic test equipment

Given we can accept automatic test equipment benefits in the long run, it remains to decide how best to use the equipment. In other words, what areas of production are best served by automatic test equipment?

There is no easy answer. On the one hand, it is important to ensure a supply of good components, so use of automatic test equipment at the goods inward stage may be beneficial. On the other hand, most faults occurring on assembled products are due to manufacturing defects, so automatic test equipment to test completed but unpackaged assemblies would seem logical, too.

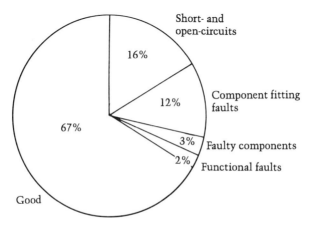

**Figure 4.12** *Typical printed circuit board faults shown as percentages of total boards after assembly and soldering*

Figure 4.12 is a pie chart of typical printed circuit board faults as percentages of total boards after assembly and soldering. While, ideally, most boards are good and work first time, the majority of faults are seen to occur in areas which are simply manufacturing defects problems: solder bridges, open circuits, missing, wrong and incorrectly inserted components, components damage during production. Figure 4.13 shows all faults as percentages of total faults. Interestingly, manufacturing defects account for some 85% of total faults, while faulty components account for only about 10%. This would seem to argue in favour of automatic test equipment at unpackaged assembly level, not at goods inwards stage.

Of course, vendor assessment of component parts (described in Chapter 2) eases requirements of automatic test equipment at goods inwards stage – if not eliminating the requirement altogether. If components are used which form part of national or international specifications systems, bought

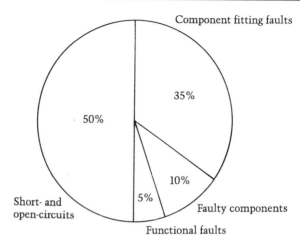

**Figure 4.13**    *Typical printed circuit board faults as percentages of total faults*

from qualified products lists from these systems, there is no requirement at all to test components as they are purchased. Testing and approval is done by the vendor, not the purchaser.

## A strategy of costs

Now, this scenario is an extremely simple one, and will doubtless make many accountants wince, but it does show how the cost of testing products may be offset, more than completely, by the advantages gained in the product's greater reliability.

The fact that defects have been reduced at an early stage in a product's life effectively eliminates the requirement for later rework of the defect. And rework is far more costly than early elimination. This principle is true at all stages of a product's life.

As a rough and oft-quoted guideline, the cost of defect correction is considered to rise by a factor of 10 with each production stage of an appliance's life. So, a defect which may cost 3p to correct at pre-assembly level (say, a faulty resistor), will cost 30p to correct at post-assembly level, £3 at packaged level, and £30 to repair in service. This factor of 10 may rise, though, as products become more and more complex.

This should be sufficient incentive to make companies doubly aware of the importance of having a good test strategy. Naturally, part of this test strategy must take into consideration all costs incurred when automatic test equipment is incorporated into an organization's test strategy (see Chapter 2). On the other hand, all benefits which accompany these costs due to the existence of automatic test equipment must equally be taken into consideration in any financial analysis of test strategy. Although this is a complex affair manufacturers, keen to promote their automatic test equipment systems, invariably are willing to help potential customers analyse test strategy costs and benefits. In the end, it should be noted that any analysis is based on predictions and is simply a guideline to the future – it is not infallible.

# 5 Test methods and processes

The aim of this chapter is to describe the various processes used by automatic test equipment systems to test a product. Naturally, the number of processes used in any particular test, and their complexity, depends totally on the product. The more complicated the product, the more processes are generally used to test it. The type of process, too, depends purely on the product under test.

To this end, most of this chapter is taken up describing a straightforward list of processes in a descriptive format. General processes are considered first, followed by digital testing processes, then analog testing processes. Where different processes are all merely sub-processes of a generic type of process they are grouped together under the generic process name.

## General processes and considerations

Tests performed in functional, in-circuit and combinational test processes depend primarily on the type of circuit being tested – digital or analog. Digital circuits, based on logic, are inherently easier to test from the viewpoint of automatic test equipment. Even at individual gate level simple yes/no tests can be performed quite easily. Analog devices, on the other hand, must be tested to ensure they fall within a broad range of acceptable performances.

A process which defines what is wrong with a circuit, part of a circuit, or component is sometimes called a **diagnosis**. This is analogous with a medical diagnosis, where a doctor or surgeon performs medical tests prior to announcing what is wrong, if anything, with a patient. In both medical and test equipment diagnosis, this process may or may not occur at the same time as finding a fault in the first place. Test equipment **fault location** often occurs first – a test identifies *something* is wrong, but not exactly *what* is wrong. However, certain types of automatic test equipment allow a diagnosis to occur as tests are carried out. Specifically, such tests are usually those performed on individual parts of a circuit – if one test on one component shows a fault, the diagnosis is often as simple as that component being faulty.

Automatic test equipment is mechanically connected to a circuit to be tested by a fixture (see Chapter 3). Electrically however, drivers, receivers and sensors are required within automatic test equipment which interface circuit and system. Drivers and receivers for digital testing, and sensors for analog testing are discussed later, when considering digital and analog test processes individually.

Tests, whether on digital or analog circuits, are only of any use if three main functions are inherently undertaken by automatic test equipment:

- Initialization – in which all parts of the circuit are set to known states before test patterns and processes are commenced. Digital initialization usually involves application of test pattern vectors, comprising parallel bit patterns. Analog initialization, on the other hand, involves application of analog parameters such as voltage, current, waveforms and so on.
- Observation – adequately monitoring a circuit. This may only be carried out if test points are available at required points in a circuit.
- Control – an automatic test equipment system can only fully carry out adequate testing if it is capable of controlling the tested circuit. Put another way, if *all* circuit functions cannot be controlled by an automatic test equipment system, a circuit cannot be *fully* tested.

These functions may be entirely separate from or form part of tests carried out by an automatic test equipment system. Initialization, observation and control of a circuit together imply the circuit is **testable**. Inevitably, whether a circuit is testable or not depends on its design. In effect, testability must be designed into a circuit in order that adequate test may be later performed. Readers are referred to Chapter 4, where **designed for testability** (DFT) is discussed, and later in this chapter where scan test (an example of an application of design for testability) is considered.

## Partitioning

One of the key concepts of testability in automatic test equipment systems is isolation of tested circuit devices or parts from the remainder of a circuit. This is known as **partitioning** and, as parts within an isolated partition are often known as a **cluster** of parts the process is often called **cluster testing**. Partitioning is extremely useful because those parts within a cluster are viewed as if they are by themselves – not in the total circuit. Tests may be performed on the cluster as if it is an isolated circuit. Thus, a cluster is checked for operation and verified, prior to consideration as part of the whole circuit.

Clusters may be as small as a single component, may contain a handful of components, or may be as big as a whole tested circuit. Where clusters may be changed by the automatic test equipment in real-time, as a test process proceeds – that is, adapting to test requirements as previously undertaken tests define – the technique is commonly called **variable in-circuit partitioning**, **multimode testing**, or **polyfunctional testing**. At the extremes, this effectively gives automatic test equipment the ability to vary itself between full in-circuit test and full functional test strategies.

Partitioning, of whatever type, relies on the ability of the automatic test equipment system to isolate a partition from its surrounding circuit. How this is done depends largely on wehther a digital circuit or an analog circuit is being tested. Methods are considered individually, later.

## Digital testing processes and considerations

An automatic test equipment system for testing digital circuits typically contains a set of parallel digital drivers, outputs of which are used to stimulate the tested circuit in a defined way. Results of these stimuli are monitored using a set of digital receivers.

### Digital drivers

Drivers are usually three-state devices; allowing high states and low states to be driven, as well as an undriven high impedance state. Often high and low state voltages are of standard TTL levels, although programmable state voltages are available. A possible driver is shown in Figure 5.1 along with its truth table.

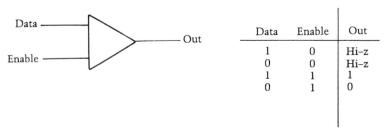

| Data | Enable | Out |
|------|--------|------|
| 1 | 0 | Hi–z |
| 0 | 0 | Hi–z |
| 1 | 1 | 1 |
| 0 | 1 | 0 |

**Figure 5.1** *Typical digital driver of a digital tester, with truth table*

A simple output waveform from such a driver is shown in Figure 5.2a, where the output is in an undefined state (one of its three possible states), then is driven high as a test process begins.

Often, however, such an output is too simple and some kind of **formatted output** is required. Figure 5.2b shows output of a **formatted**

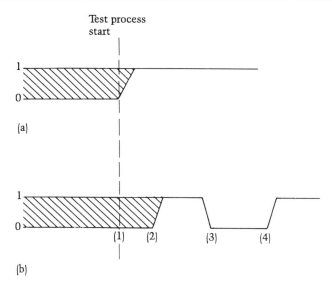

Test process
start

(a)

(1)  (2)  (3)  (4)

(b)

**Figure 5.2** *Digital driver output waveforms (a) simple output (b) formatted output in a surround by complement to zero format*

**driver** where a number of distinct stages occur in output signal: (1) the test process begins, at which time driver output is in an undefined state; (2) output is driven high; (3) output is driven low; (4) output returns high. Effectively the driver's output is modulated by data from the automatic test equipment system's controlling program. This example shows modulated data in a surround by complement to zero (SBC-0) format. Other format alternatives include **surround by complement to one** (SBC-1), **surround by high impedance** (SBZ), **return to one** (RTO), **return to zero** (RTZ), **and non-return to zero** (NRZ).

Formatted drivers are used in automatic test equipment systems to impose stresses on a circuit's input timing specification. To this end, driver formats of an automatic test equipment system are often programmable and switchable by the controlling program. To further enhance this facility automatic test equipment systems often allow rise and fall times of drivers to be specified under program control. Thus, formats and rise/fall times may be varied either before or during tests.

Imposition of timing stresses has a dual function. First, devices in the tested circuit are stressed to their timing limits. Second, these stresses are imposed at a much lower speed (hence a much lower clock rate is used) than would normally be possible.

Where a fixed logical state output is required of a driver, pull-up or pull-down resistors are usually provided, switched to suit high state or low state.

## Digital receivers

Digital receivers used in automatic test equipment systems are combinations of basic comparators (of an analog nature, incidentally) coupled with logic gates to determine whether received data is as expected. Figure 5.3a shows such a simple receiver. A comparator compares the received data voltage with a defined reference voltage, giving a result which is combined with the expected data in an exclusive OR gate. When data received is the same as data expected the gate output is low. However, when received and expected data are not the same, the gate output is high, indicating an error.

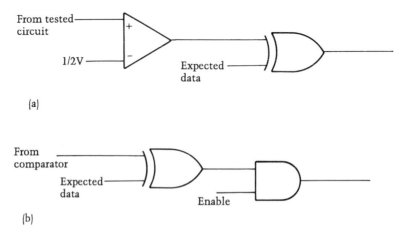

**Figure 5.3**  *Typical digital receiver of a digital tester (a) basic receiver (b) additional circuit to allow definition of window during which receivers respond*

Error results may be used in several ways – generally depending on automatic test equipment system designer's ingenuity. For example, by further gating the error signal from the exclusive OR gate, as shown in Figure 5.3b, it is possible to define one or more windows during which a circuit's response to a driven stimulus is sampled. Then, by varying sampling delays and periods under program control, circuit response times such as data setup time, data hold time, propagation delays and so on may be measured.

## Interfacing

Drivers and receivers are usually interfaced with a tested circuit via a multiplexing arrangement, illustrated in Figure 5.4. As connections between automatic test equipment and tested circuit occur through a

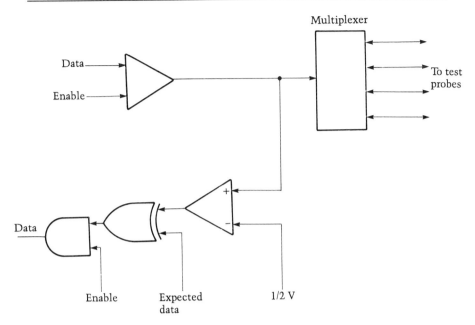

**Figure 5.4** *Multiplexing digital tester outputs and inputs to allow access to more tested circuit nodes*

mechanical fixture (see Chapter 3) and as such fixtures usually feature test probe pins, terms such as **multiplexing of pins**, and even **multiplexed pin architecture** are used. However, it is important to remember pins themselves are not multiplexed, merely internal connections to drivers and receivers.

The reason for multiplexing driver and receiver connections is to allow a greater number of test probe access points (that is, pins) onto the tested board. So, automatic test equipment with say, 100 driver/receivers may have a **multiplexing ratio of 4:1**, allowing up to 400 test probes. Effectively each driver/receiver has four test probes to stimulate and monitor. Sequentially switching between test probes is therefore necessary.

Such multiplexing can give automatic test equipment a cost advantage when compared with others having a lower multiplexing ratio. However, multiplexed automatic test equipment may not allow a sufficiently large number of test points for a given application to be measured simulta-neously – in real-time. Following the example just given, of 400 test probes multiplexed to 100 driver/receivers, a single integrated circuit with over 100 pins simply cannot be monitored in real-time.

Switching between driver/receivers and test probes is usually performed with electromechanical relays, which feature a very low on-resistance.

Relays, however, place a limiting factor on multiplexing rate. Semi-conductor switches, which could operate at significantly higher rates, have a much higher on-resistance, which would affect measurements taken, so are not often used. Some automatic test equipment, however, features both relay and semiconductor multiplexing. Here, relays are used for accurate, but slow, measurements; while semiconductor switches are used for simple measurements (such as open and short tests) where accuracy is not required but speed is.

### Electrical isolation

Where individual devices or parts in a tested digital circuit are to be tested they may be effectively electrically isolated from the remainder of the circuit, for short periods of time at least, by forcing outputs of preceding circuit devices to required logic states. If precautions are taken this may be done with no damage.

Circuit devices are effectively overwhelmed by this technique, so control of following circuit parts may be assumed by the automatic test equipment system, rather than by the tested circuit. Inputs to the selected circuit part are controlled solely by the automatic test equipment system and, in essence, the circuit part acts as if it is disconnected from the whole circuit. It is this concept which allows in-circuit and combinational testers to partition tested circuits into clusters or individual devices.

Forcing outputs in such a process is known as **overdriving** and two different nodes of operation are necessary. First, if an output is high and is forced low the process is called **nodeforcing**. Second, a low output forced high is known as **backdriving**. Component loading when nodeforcing is considerably lower than when backdriving. Figure 5.5a illustrates the use of nodeforcing and backdriving, where two of an integrated circuit's output stages are overdriven by signals from automatic test equipment. Graphs of nodeforced and backdriven signals are compared in Figure 5.5b. Current out of an output stage during nodeforcing is seen to be much lower than current in during backdriving. Nodeforcing is made further irrelevant when you consider its effect: forcing a high output low is the same as simply applying a short circuit on the output, which most if not all modern integrated circuit devices are designed to handle, anyway.

Overdriving preceding circuit parts to control a following circuit part does not always, however, guarantee absolute control. Where the output of the controlled circuit part is common with other circuit part outputs, those other circuit parts may affect operation of following circuit parts, so must be controlled too. Such a situation often arises where digital feedback loops around counters are in a circuit. Typically, **digital guarding** is used, simply injecting fixed logic states at vital points in the circuit part, to stabilize following circuit parts thus ensuring a known state.

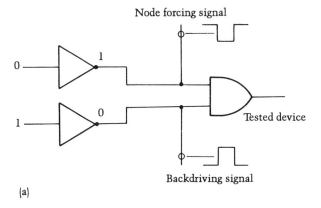

(a)

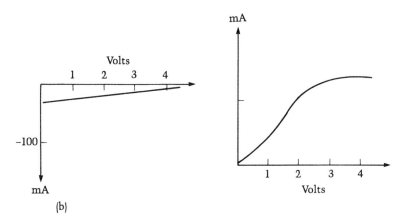

(b)

**Figure 5.5** *Electrical isolation of tested circuit parts by overdriving (a) nodeforcing and backdriving (b) graphs of nodeforced and backdriven currents*

Overdriving is carried out using high current pulses of low energy and short duration. Two specific types of pulses are used: short pulses to overdrive data bits, clocks and reset signals; longer pulses to overdrive device enable, set and clear signals.

As pulses of opposite logic levels to those which may be otherwise expected in a device output are used, it is reasonable to expect that damage may occur to a device's output stages. There are three main causes for concern regarding damage by overdrive:

- Localized heating effects in output stage transistors causing secondary breakdown.

- Localized heating effects in metal conductors – specifically bond-wires between die and output pin connections – which may cause failure if the temperature is too high.
- Voltage overshoot in CMOS devices, causing outputs to latch into a permanent logic state. This is more a problem of automatic test equipment hardware design than it is a concern for overdriving devices.

Of these, bond-wire failure due to excessive temperatures is the most significant problem, and is thus the limiting factor when considering use of overdriving.

Bond-wire failure characteristics vary according to integrated circuit package. Two main package-types, ceramic and plastic, use different metals for bond-wire connections and placements. Plastic packages use gold wires encased in the package epoxy. Ceramic packages, on the other hand, use bond-wires made of mainly aluminium (99% aluminium, 1% silicon) in a gaseous medium. Naturally, heating characteristics of the types differ, as do maximum safe temperatures before failure occurs.

Failure is further complicated because, where long backdriven pulses are used – say, over 1 ms or so – or where more than one output of a device is backdriven, the temperature of the die itself rises significantly due to thermal conduction along the bond-wires, with a consequent increase in bond-wire temperature.

Aluminium bond-wires exhibit relatively poor thermal conduction and thus have greater temperature rises in bond-wires themselves, exhibiting temperature peaks at specific points along their lengths. Gold bond-wires encased in plastic epoxy, on the other hand, exhibit much greater thermal conduction, so generally remain at more even temperatures without peaks.

Although it is initially logical to assume that maximum bond-wire temperature is limited by bond-wire melting-point, this is not the case. Aluminium bond-wires exhibit embrittlement if maintained at a high enough temperature (something over 200°C) for too long. Thus a practical maximum temperature at any point along a ceramically packaged device's aluminium bond-wire is 200°C. Gold bond-wires packaged in plastic epoxy have a lower maximum practical temperature limit (around 125°C), imposed by stress on bond-wires if the plastic encasing them reaches its glass transition temperature causing the plastic to shrink away, effectively stretching bond-wires until they break.

Aluminium's relatively poor thermal conduction with resultant temperature peaks, against plastic's comparatively low glass transition temperature limit, mean that overdrive current and time limits for both types of device are similar. Although this simplifies automatic test equipment design somewhat, responsibility for ensuring overdrive pulse currents and pulse times are not exceeded still remains the province of automatic test equipment.

## Scan testing

Another method of isolating devices or parts of a circuit while testing them is **scan testing**, which is a recent innovatory attempt to allow high-density circuits to be fully tested. Described fully as a test strategy in Chapter 4, scan testing is a means of looking at inputs and outputs of devices in a circuit by using a special scan test mode, which effectively stops circuit operation and allows specific test states to be serially entered prior to running the circuit for a known number of cycles. Similarly, all states may then be serially extracted for examination.

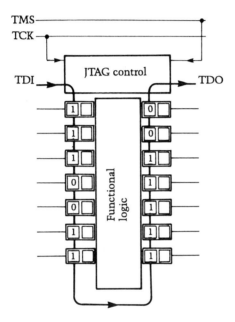

**Figure 5.6** *Scan test integrated circuit showing four-wire testability bus (a reprint of Figure 4.7)*

The basis of scan testing is the Joint Test Action Group's (JTAG) proposal to implement certain features inside integrated circuit devices at manufacture stage. These proposals are now incorporated in the American proposed standard IEEE P1149.1. Here integrated circuit devices feature scan test control and input/output latches extra to conventional functional circuits (Figure 5.6 – a reprint of Figure 4.7). Such devices are commonly known as **scan design** devices.

Scan design devices all feature a four-wire bus, known as the **four-wire testability bus**, which allows data to be entered and extracted to and from the input/output latches quite simply, in a four-stage process:

- **Test–mode select** (TMS) signal is set so the device is in test mode, in which all input/output latches are connected together to form a shift register, known as a **scan register**, while all normal inputs and outputs to and from the devices through the latches are suspended.
- A test vector is serially entered into the scan register, via the **test data in** (TDI) connection. Clocking for this operation is provided by applying a clock signal to the **test clock** (TCK) connection.
- **Test–mode select** signal is reset so the device is in normal mode. A defined number of cycles of operation (usually one) are stepped through.
- Resultant vector in the scan register is removed serially from the **test data out** (TDO) connection, again applying a clock signal to the **test clock** connection and examined.

Scan test operation using such a device gives rise to two important features. First, as scan registers are simply shift registers, test vectors clocked into a scan register cause corresponding vectors to be clocked out. Thus input and output of vectors is a single concurrent feature, rather than two isolated features suggested by the four-stage process discussed here.

Second, and more important, vectors from one device may be entered into a subsequent device (by connecting *test data outputs* of devices to *test data inputs* of following devices), allowing a single, large scan register to be built-up comprising scan registers of every device in a circuit. A single test vector, of sufficient bits to give the required input to every device, may then be locked in via the *test data in* connection of the first device. Circuits featuring this are said to be of **boundary scan design**, and the complete scan register built-up with individual devices' scan registers is known as a **boundary scan register**.

### Virtual channel testing

One of the problems arising in the use of automatic test equipment to test high-density circuits occurs simply because of the high-density. Where very large scale integrated circuits, particularly of surface mounted design, are used in complex circuits it is often impossible for mechanical fixtures to access sufficiently large numbers of test points in a circuit to allow adequate testing. Even if high-density circuits are not used, board coatings, such as conformal coatings, may deny test point probe access.

Scan test procedures in which all devices in a circuit are of scan test design give an eventual solution but, for the foreseeable future at least, circuit designs are rarely going to be totally of scan test design. As such, devices and components of non–scan test design are likely to be used with scan test design devices in circuits. Typically, non–scan test devices form clusters within scan test design device boundaries.

This is a test problem which, fortunately, scan test processes seem

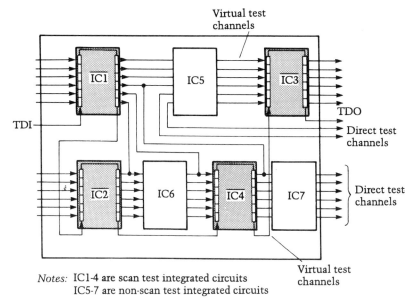

**Figure 5.7** *Printed circuit boards may contain scan test and non-scan test components. At the junctions between, virtual test channels are available (a reprint of Figure 4.8)*

sufficiently adaptable to sidestep, it not overcome. Figure 5.7 (which is a reprint of Figure 4.8) shows a possible circuit, in which scan test design devices are positioned around a cluster of non-scan test design devices. Mechanical access of test point probes in or around the non-scan test design device cluster is denied, so at first sight testing of this part of the circuit is impossible. However, the non-scan test design device cluster is effectively surrounded by test points which are commoned with input or outputs of surrounding scan test design devices. By using scan test features, clocking out test vectors from the boundary scan register of the circuit, logic states at those commoned points may be accessed. Commoned test points used in this way are known as **virtual test channels**.

Cluster size has a great bearing on how effective virtual channel testing is. Obviously, the smaller a cluster, the more closely it may be tested. Careful circuit design, using non-scan test design devices in prudent circuit positions, ensures cluster size is as small as possible. Ultimately, single non-scan test devices between scan test design devices allows total test point access to individual devices by virtual channel testing processes. More likely, non-scan test design device clusters are likely to hold a handful of devices. With sensible application automatic test equipment processes will give adequate fault diagnosis for mixed scan test and non-scan test design circuits.

## Digital circuits

There are four main types of digital circuit:

- Combinational logic – where outputs are dependent on inputs at any one time.
- Sequential logic – where outputs depend on present and previous inputs and outputs. Sequential logic circuits are characterized by memory or feedback.
- Bus-based logic – combinational and sequential logic circuits together, where components are interconnected with some form of bus.
- Random logic – combinational and sequential circuits together, where components are not interconnected by a bus.

These digital circuit types effectively mirror the types of tests automatic test equipment systems must use.

### Combinational logic

Testing combinational logic is a relatively simple job of applying digital inputs and observing circuit outputs. Ideally, all possible variations of input changes must be applied, while all consequent output variations must be considered. However, such **exhaustive testing** is rarely economically possible, in terms of pure cost and time. A circuit with ten inputs, for example, has $2^{10}$ possible input variations, calling for a consequent $2^{10}$ separate tests to be performed for exhaustive testing to be carried out.

Further examples of the uneconomic strategy of exhaustive testing can be seen when considering microprocessors. Motorola's MC6800 micro-processor, it has been calculated, would take some two million years to execute all instructions for all possible combinations of input data and internal states at typical test rates. A 32-bit microprocessor similarly exhaustively tested would require a test time greater than the known age of the universe.

Initialization, observation and control of combinational logic circuits is quite simple. A test pattern vector of inputs initializes the circuit, setting all parts to their predetermined states. Further test pattern vectors are applied in a number of test processes, while observation is simply a matter of observing signals from an adequate number of test points. Where individual devices in the circuit are tested, test pattern vectors are applied to device inputs via test points, while outputs are observed after a time offset. Such **bit pattern measurement** is common to most automatic test equipment systems. All input and output states are usually held in system memory as a truth table for a particular type of device. A deviation from expected bit pattern output indicates a faulty device.

It is possible to define correct response as a **signature**, comprising a single number in a test process known as **signature analysis**. In a combinational logic circuit, the bit pattern of a specific correct output of the circuit or a device within it may be classed as a signature if all other variations of output are incorrect and produce other bit patterns.

Finally, control is inherent in initialization test patterns (combinational logic produces controlled outputs entirely dependent on inputs).

## Sequential logic

Sequential logic circuits are more difficult to test, because faults within a circuit are not necessarily detectable at its output. To test for such faults automatic test equipment systems must, ideally, allow:

- Adequate numbers of test points. Where possible, test processes must provide stimuli in the form of test patterns to transmit a fault condition (or its digital consequence, at least) through the circuit – known as **walking out** – to a test point. Obviously sufficient test points must be provided in a circuit to allow this.
- A means of isolating individual components within a circuit. Complete testing of a component cannot be undertaken without some means of electrically isolating it from the remainder of the circuit.
- Access to memory devices' reset and preset lines.

While in–circuit testing inherently allows these scenarios, functional testing rarely does.

Initialization of sequential logic must eliminate all unknown states occurring after power-up, before tests can be carried out. This is done by walking out the unknown states in a defined procedure of applied inputs. Fixed sequential procedures to do this are effectively a set of test pattern vectors applied after every power-up situation, and are known as **synchronizing sequences**. It is possible, however, to devise initialization procedures which differ according to circuit behaviour, adapting to logic states at test points. Such procedures are known as **homing sequences** or **adaptive homing sequences** (AHS).

Automatic test equipment control of sequential logic circuits has to take into consideration all possible states of the circuit, throughout its whole sequence of operating states.

Test processes used for sequential logic circuit testing vary considerably. Initially, an automatic test equipment system must simulate and measure test patterns likely to occur in use, under reproducible and real-time conditions. A high degree of synchronization between patterns and signals generated by the automatic test equipment system is required. The only way to do this effectively is with a master high-speed clock, divided down to all required rates.

Sequential **signature analysis** is common, in which a sequential response from a circuit is compressed into a signature. Often a cyclic redundancy check (CRC) is used to compress a long stream of bits received into a single signature, which is compared with an expected signature. Signature analysis has a couple of drawbacks, however. First, an incorrect signature received merely indicates the presence of one or more faults; it neither shows the fault, nor when it occurred. Second, all bits of the response from the circuit must be received and compressed into a signature before signature analysis may occur; faults cannot be detected mid-stream. On the other hand, signature analysis has the advantage that it may be used to indicate time related errors (even though it cannot show where or when they occur). It is regularly used, therefore, to test sequential logic circuits and, indeed, bus-based and random logic circuits.

## Bus-based logic

Where digital circuits are constructed around a bus system, with microprocessors, addresses and data, initialization, observation and control of circuits becomes quite complex. There are a number of processes which have been developed specifically for these purposes, however. As these are all important processes they are covered individually in following sections.

### Bus timing emulation

Although originally developed by *Columbia Automation* (now *Zehntel Performance Systems*) as an aid to microprocessor-based system design, **bus timing emulation**, sometimes known as **bus-cycle emulation**, quickly became adopted as the basis of a process of automatic test equipment which allows complete simulation of a tested bus-based system, without the system's microprocessor. Some manufacturers even go so far as to use bus-timing emulation as a complete test strategy – performance test – in its own right.

In bus-timing emulation a microprocessor is usually put into a hold or reset state by the automatic test equipment, although it may be physically absent if a test strategy requires. Absence is most often used if circuits are tested on manufacture prior to microprocessor insertion, but situations may arise in which microprocessors must be removed before a circuit is to be tested.

Without an on-board microprocessor to maintain control, automatic test equipment itself must issue sequences of vector patterns which the microprocessor would otherwise perform. Thus the automatic test equipment effectively exercises remaining parts of the tested circuit, ensuring correct operation of the whole circuit. Note, however, the microprocessor itself is not tested and this must be done at a later test stage.

Automatic test equipment using bus timing emulation has a timing emulator which replaces the tested system's microprocessor. It may be synchronized with the tested system's clock and itself produces synchronous timing signals so that a whole system may be totally controlled by the automatic test equipment. Key to the test process is use of a high level programming language, which allows simple programmed control of a microprocessor's complex operational procedures.

A tested system's sub-circuits are functionally grouped into basic bus-cycle types – read out of memory, write into memory, input, output, opcode fetch, interrupt acknowledge – and corresponding sections of the test process allow the automatic test equipment to apply and measure data within each group. Dynamic (in real-time) or static (step by step) testing of tested systems may therefore be undertaken. This means not only are the microprocessor's normal operational procedures within the tested system tested, but extra procedures may be used by the automatic test equipment to test the system more extensively – procedures which may even be beyond the microprocessor's capabilities.

Many microprocessors allow electrical isolation of inputs and outputs with three-state drivers and receivers so it is a simple job of driving them into the high impedance state, which effectively isolates devices from the bus, while the automatic test equipment system emulates bus timing operation. Alternatively, systems may be tested prior to insertion of microprocessors, or electrical isolation techniques (see later) may be used.

## Memory emulation

Automatic test equipment may control a bus-based system by instructing the systems' microprocessor to address memory within the automatic test equipment system, as if it is its own. As a process this is known as **memory emulation** and a common name for this form of automatic test equipment operation is **capture mode**. Effectively, the microprocessor's own memory is masked from the microprocessor. If the automatic test equipment system memory holds programs, known as **target routines**, microprocessors may be totally controlled and tested, along with system operation.

Target routines are of two main types: **diagnostic test routines** (DTR) which test functional blocks within the tested system, and **idle routines** which keep the microprocessor (and hence the tested system) in a known condition during changeover between tests.

Under memory emulation a microprocessor retains control of its own system (bar memory) while allowing automatic test equipment system control of its operation. Selective dynamic (real-time) and static (step by step) tests of a complete system are possible.

Microprocessor memory buses must be controlled by the automatic test equipment system, such that its own memory appears invisible to it. Selective electrical isolation of a microprocessor's inputs and outputs by forcing them to determined logic states (see later) is the only way this may be effected. Isolating by using the high impedance third state does not allow correct microprocessor operation and so is not possible.

## Analog testing and considerations

Analog testing techniques require broadly at least, similar considerations as digital testing techniques.

### Analog sensors

Analog tests are normally performed using a sensor arrangement based around an operational amplifier (although impedance bridge techniques are occasionally used). Generally, the non-inverting amplifier input is connected directly to ground potential, shown in Figure 5.8, where feedback from output to inverting input exists, which means the operational amplifier amplifies in inverting mode. In such a circuit, given ideal conditions, two operational amplifier properties are useful:

● Input resistance is infinite – put another way, no current flows into the operational amplifier, therefore output voltage $V_0$ is independent of output load. Instead, current from the tested circuit $I$ flows through the feedback resistor $R_f$.
● Voltages at each operational amplifier input are equal; the inverting input is said to be at **virtual earth**.

The output voltage of the circuit is given by:

$$V_0 = I \times R_f$$

Thus measurement of output voltage gives a direct indication of tested circuit analog parameters.

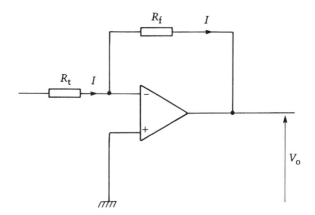

**Figure 5.8**   *An analog sensor circuit, based on an operational amplifier*

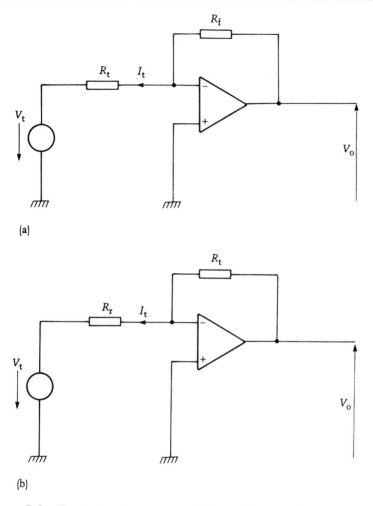

**Figure 5.9**  *Typical analog sensors (a) applying a voltage across the component to be measured (b) forcing a current through the current to be measured*

Two test circuits are derived from that in Figure 5.8, in which the tested component is located in either the measurement amplifier's input circuit or feedback circuit, shown in Figure 5.9a and 5.9b.

*Voltage forcing*
Figure 5.9a shows a circuit in which a resistor $R_t$ is tested by applying a voltage $V_t$ across it, which by Ohm's law creates a current $I_t$ through it:

$$I_t = V_t/R_t$$

which causes an output voltage from the measurement amplifier circuit:

$$V_o = I_t \times R_f = \frac{V_t \times R_f}{R_t}$$

So, the tested resistor value is:

$$R_t = \frac{V_t \times R_f}{V_o}$$

Inverting this gives:

$$\frac{1}{R_t} = \frac{V_o}{V_t \times R_f}$$

which shows the measured output voltage to be proportional to the tested resistor's admittance, so that this measurement circuit is commonly called an **admittance measurement** circuit. Sometimes, because a known voltage is applied across the tested resistor, it is called a **voltage forcing measurement** circuit.

### Current forcing

Figure 5.9b shows a similar circuit, in which a resistor $R_t$ is positioned in the measurement amplifier's feedback loop. A voltage $V_t$ applied to a reference resistor $R_r$ causes a current $I_t$ through the feedback resistor, giving an output voltage: $V_o = I_t \times R_t$.

But, as $I_t = V_t/R_r$, then:

$$V_o = \frac{V_t \times R_t}{R_r}$$

and so the tested resistor value is:

$$R_t = \frac{V_o \times R_r}{V_t}$$

As the measured output voltage is effectively proportional to the impedance of the tested components, this measurement circuit is commonly known as an **impedance measurement** circuit. Sometimes, because a known current is forced through a known resistor, it is called a **current forcing** measurement circuit.

### Impedance measurement

Though shown so far only measuring resistors, these two measurement circuits are adapted to give measurement of capacitors and inductors by using an AC voltage source $V_t$. A measurement of AC output voltage with an AC voltmeter gives simple results but, for accuracy, values are often measured using sampling circuits to measure current through and voltage across the tested impedance.

The principle is extended if samples are taken at two points during the AC source frequency (voltage maximum, then $90°$ later), from which all parameter values of the tested impedance may be calculated. Such a circuit is known as a **true-phase four quadrant measurement** circuit, sometimes a **quadrature detector**, and is capable of determining, individually, components connected in parallel to form a complex impedance.

## Electrical isolation

In simple **two terminal measurements**, sometimes known as a **two-wire measurements**, shown so far, it is usual to apply a voltage to a tested impedance and to measure the resultant current through it using an operational amplifier. Accuracy of measurement relies totally on there being no additional current into or out of the operational amplifier measurement circuit. Where components are measured in isolation out of circuit this is possible, but in most cases components are in circuit – so other voltages and currents exist which will cause incorrect measurements of component values.

Circuits used in automatic test equipment systems to counteract other in-circuit voltages and currents, effectively allowing individual components within a circuit to be electrically isolated from the remainder, rely on a principle of nulling currents around the tested component by connecting all surrounding nodes to the same potential, allowing the measurement circuit to make an accurate measurement. There are many forms of such isolating circuits, all known as **guarding circuits**.

The simplest guarding circuit is shown in Figure 5.10 to illustrate the guarding principle. Tested resistor $R_t$ is shown in parallel with two other resistors $R_1$ and $R_2$, together forming a network. Effectively, there are three nodes in the tested network: A, known as the **force node** or the **stimulus node** as voltage from the automatic test equipment internal measurement circuit is *forced* onto or *stimulates* that point; B, known as the **sense node** or the **measurement node** as current through the test resistor $R_t$ is sensed from or measured at that point; C, known as the **guard node** as the **guard** earth voltage is applied to that point.

Node B is connected to the inverting input of the operational amplifier, so is at virtual earth. Node C is connected directly to earth. Consequently, in theory at least, no current flows through resistor $R_1$. Node A is connected to the measurement circuit's voltage source so current through the tested resistor $R_t$ is sensed by the operational amplifier, giving a true measurement within limits (see later). Current through resistor $R_2$ does not affect measurement accuracy.

As this circuit has three connections between the measurement circuit

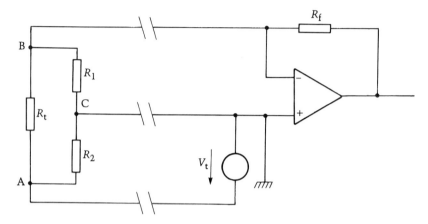

**Figure 5.10**   *Guarding illustrated as a principle. By nulling currents in nodes around the component to be measured, components may be measured in-circuit*

and the tested component circuit, it is known as a **three terminal measurement** circuit or **three–wire measurement** circuit.

Limits imposed on measurement accuracy are defined by what is known as **guard ratio**, where the guard ratio is the ratio between current flowing in the guard path (between ground node and sense node – through what is termed the **interfering resistor**; resistor $R_1$ in this example) to current flowing in the measurement path (between force node and sense node – through the tested resistor). From this definition, it does not take a genius to calculate that the guard ratio is given simply by the ratio of tested resistor value to interfering resistor value ($R_t : R_1$).

Three-terminal measurement circuits are only accurate with a guard ratio of no higher than about 100:1, which means a tested component with an interfering component of less than about a hundredth of its value cannot be measured accurately. Thus if an interfering resistor has a value of 100 $\Omega$, resistors of more than 10 k$\Omega$ cannot be accurately measured.

Where higher guard ratios are required of measurement circuits, the answer is usually to resort to additional and separate sense connections at tested network nodes – to allow compensation of voltages dropped along one or more of force, sense and guard node connections. To this end, **four terminal measurement** also known as **four–wire measurement** uses a separate sense connection to the guard node. Similarly, **five terminal measurement** also known as **five–wire measurement** has separate sense connections to force node and sense node. Finally, **six terminal measurement** also known as **six–wire measurement** uses separate sense connections to each of the three nodes.

Six terminal measurement gives guard ratios of 10000:1, much improved over that of three terminal measurement. However, every node participating in six terminal measurement requires two operational amplifier measurement circuits, and two test point probes.

## Auto-delay, auto-dwell

Generally, measurement values are not stable until some time after stimulation. Consequently, measurement results are inaccurate unless values have stabilized. One answer is to fix a time after which all values can be assumed to have stabilized, only then taking measurements. A disadvantage of this technique is that values which stabilize sooner than this cannot be measured until the time has passed.

**Auto-delay**, sometimes called **auto-dwell** techniques take measurements at a number of intervals, comparing values with previously taken measurements. Automatic test equipment featuring such a technique only accepts a result as a final value when steps between results are within a predetermined limit.

## Measurement averaging

One of the main disadvantages with analog techniques is the well-documented problem of interference by noise of many types. Often automatic test equipment features the ability to average readings taken, over a predetermined number of measurements and time intervals, thus allowing noise cancelling.

# 6  Computer and instrument buses

An automatic test equipment system is essentially a computer system. Peripheral instruments are computer-controlled, while data passes between peripherals and computer usually along a small number of data buses – one bus for each type of data being passed. As a rule-of-thumb, where these buses are combined as, say, a single multi-cored cable, plurality is dropped and these buses are referred to as a **bus**.

Over recent years many types of data bus have been used to create automatic test equipment, all of them having different features which suit different applications. In some cases manufacturers designed a data bus unique to their own instruments; consequently restricting users' choice of peripheral instruments. Many data buses were borrowed from other electronic fields, such as the V24 (based on the earlier IEEE: EIA232) standard serial data communications bus, and the S100, 6800, and Z80 parallel microprocessor communications buses. This situation naturally led to a general non-conformity between automatic test equipment which deterred potential users; equipment from one manufacturer could rarely be used alongside another manufacturer's equipment in the same automatic test equipment system.

In 1965, however, a data bus was defined by Hewlett Packard (the **Hewlett Packard interface bus** – HP-IB) specifically for the purpose of interfacing programmable measuring instruments, accessories and a computer into an automatic test equipment system. It was a more-or-less comprehensive bus which allowed all types of programmable measuring instruments to be connected into a single automatic system, so many other manufacturers incorporated it into their instruments during the early 1970s. For the first time, users could purchase instruments from many manufacturers and be certain they could be successfully interfaced.

American Institute of Electrical and Electronic Engineers adopted the interface into its standard, the **IEEE 488**, ten years after it was first developed and American National Standards Institute incorporated it as the standard **ANSI MC 1.1**. Finally, International Electrotechnical Commission adopted it in standard **IEC 625** (**BS6146** in the UK), defining a different connector plug. All standards are fundamentally compatible so it became a worldwide instrument bus standard, referred to by any of these standard numbers or, popularly, a new name which avoids any ambiguity; the **general-purpose interface bus** (GPIB), discussed in depth in Chapter 7.

Adoption of a GPIB standard effectively revolutionized automatic test equipment. Most automatic test instrument manufacturers produce instruments which have an interface to the GPIB, which means the instruments are programmable and may be used by themselves to measure and display readings of a system under test, or as part of an automatic test equipment system controlled by a computer.

GPIB does have its limitations, on the other hand. First, system data speed is a maximum of around 1 Mbyte s$^{-1}$, although in most applications much lower than this because overall speed depends on the slowest instrument connected to the bus. Second, a measurement system using the GPIB may comprise many peripheral instruments, all taking considerable workbench area.

Despite its limitations, though, such a standard at least made instrument manufacturers realize it is possible to interface peripheral instruments together to construct systems specifically suiting varied applications which users need. It only requires to take the concept one stage further, developing a bus which allows modular instruments from any manufacturer to be added to an instrument bus within a single housing. In this way the two disadvantages of the GPIB: speed and size, are overcome.

This step began in the late 1970s when Motorola defined a computer bus, which it called **Versabus**. Intended specifically for Motorola's 68000 microprocessor this bus has very little use as an instrument bus in automatic test equipment systems, where an abundance of microprocessors appear. This prompted Motorola, with Mostek and Signetics, to define another bus along similar lines in 1981 – VMEbus – short for *versa module eurocard bus*.

Although based on Versabus, VMEbus is independent of any particular microprocessor and so forms a better basis – though not ideal – for automatic test equipment system manufacture. Also, as its name suggests, a Eurocard format allows equipment to be manufactured on a modular plug-in basis. VMEbus is now standardized in IEC821 and ANSI IEEE1014, and is described more fully, along with its improvements and upgrades in Chapter 8.

The next step was taken in 1987 by a consortium of test equipment manufacturers comprising Colorado Data Systems, Hewlett Packard, Racal-Dana, Tektronix and Wavetek. Realizing the potential of such an instrument bus, the consortium proposed adoption of an existing computer bus (VMEbus, as it happens), upgrading and extending it to include specifications of module size and performance. Result; VXIbus (short for *VMEbus extensions for instrumentation*) forms the basis of complete, high-speed, automatic test equipment systems in modular form. VXIbus data transfer rates are in the region of 1 Gbyte s$^{-1}$, and modules are in a range of standard sizes. Looking to avoid incompatibility with existing equipment where possible, the bus is designed to allow fairly simple interface modules to be used to connect GPIB or VMEbus instruments. Since the initial proposal, over 100 manufacturers have registered with the consortium and so are licensed to build equipment to use with VXIbus.

For the next few years, at least, VXIbus looks set to become the basis of most automatic test equipment systems. Apart from advantages of size and speed over other bus systems, it has been quoted that an automatic test equipment system using VXIbus will be around one-third the price of a similar system using one of the other buses and about one-tenth the size. VXIbus is discussed in depth, in Chapter 9.

## Buses

This chapter is not specifically about the general-purpose interface bus, VMEbus or VXIbus – separate chapters are devoted to each. Instead many other buses are compared here, in an attempt to discuss computer buses in a general form, with respect to automatic test equipment systems. Suitability of any particular computer bus can thus be considered, along with reasons for its use in any particular application.

A bus links parts of a computer system with a common electrical highway. How many parts are linked depends on the bus, on the component parts, and on the electrical interfaces within each. Some buses may be no more than a handful of copper tracks between two integrated circuits on a printed circuit board. Other buses may be complete systems, incorporating electrical interfaces, power supplies, chassis and so on, all hardwired to connectors into which many printed circuit boards plug in. In this respect, complexity of a computer bus depends mainly on the data communication levels required of it.

### Bus data communications levels

Inside any computer system there are several levels at which bus communications can occur. Considering a range of systems, the main levels are:

- Between integrated circuits on a printed circuit board.
- Between printed circuit boards in an instrument.
- Between instruments.

- Between a controller and peripheral instruments.
- Between separate systems, each with its own controller.

These levels are worth bearing in mind as we look at buses used in automatic test equipment systems. Understanding the levels of data communications taking place within bus systems will help us to understand the capabilities of each.

Looking at bus data communications levels like this, allows us to consider buses as having three very different bus co-existent functional criteria. First, *conceptually* they must have parts which control the bus, parts which are controlled via the bus and can do nothing without instructions from controllers, and parts which (for the most part) can follow their own routines without bus control although occasional prompting may be required.

Second, data, information, commands, messages and signals must pass over the bus in a coordinated and structured manner. *Logically*, therefore, all must be defined so all parts of the computer system know what they all mean. Transmissions of all natures must be defined so all system parts may use them.

Third, *physically* all parts must fit together. Dimensions must be correct, connections to connectors must be correct, electrical voltages must be correct.

Obviously, to ensure computer bus systems fulfill all these criteria, standards are produced which define conceptual, logical and physical matters. Often these standards are produced, in the first instance, by manufacturers wishing to create a new product for reasons of profit. Sometimes manufacturers do this work alone, sometimes in a team of like-minded colleagues. Inevitably, if a standard becomes sufficiently well used, it is incorporated as a national, regional, or international standard. The general-purpose interface bus, VMEbus and VXIbus are all prime examples of this procedure. On occasion, national standards from different countries are different in name alone, and so are all compatible (a case in point: British standard BS6146 is compatible with International Electro-technical Commission standard IEC625 which, in turn, is compatible with American National Standards Institute standard IEEE488; they all define the general-purpose interface bus for programmable automatic test equipment systems).

## Advantages of buses

When producing any computer-based system there are three main developmental routes:

- Design, develop, manufacture from first principles – a custom-built system.

- Convert a standard computer – an adapted system.
- Use a modular bus – an integrated system.

Naturally, cost and ease of use are factors defining which of these three routes is chosen. A custom-built system is expensive to develop, but allows mass-product relatively cheaply. Systems are purpose-built and so perform tasks which are specifically desired, rather than tasks which may be merely adaptations of other non-specific tasks.

An adapted system is cheap to develop (the computer itself, usually a personal computer, is already developed) and, for small numbers of required new systems, may give an overall cheaper product than other routes. Systems may be inflexible, however.

Integrated systems are effectively compromises between custom-built and adapted systems, giving the best of both worlds. Development is quite cheap (the bus is already developed), and both small or medium–high production numbers are quite cost-effective.

The main advantage of an integrated computer system is its basic feature of being a standard connection device – whatever the bus connects may be replaced, renewed, moved around, reconfigured and so on, but the bus itself remains the same. In effect, it allows modularity. Integrated circuits, printed circuit boards, complete chassis-based test instruments may be unplugged and changed round without needing to change the bus.

This modularity in itself is important in two ways. First, systems may be upgraded by changing a specific module. Second, systems differ from one another merely by having different modules. So, as long as the bus itself is sufficiently well designed to cater for future needs, it is *future-proof*. Only one bus is required; whatever you want it to do in the future is accomplished merely by adding or changing modules.

## Available buses

There are many buses which are used as the basis for computer-based systems. Some of these are:

- V24/EIA232.
- EIA422/EIA423.
- EIA449.
- S100.
- Multibus.
- PC expansion bus.
- STEbus.
- GPIB.
- VMEbus.
- VXIbus.

**Table 6.1**   Comparison of a range of buses used in computer and instrument bus systems

| Bus | Level of data communications (between....) | Address bits | Data bits | Maximum speed (byte s$^{-1}$) | Range (m) | Circuit board size (mm) | Connector | Standards |
|---|---|---|---|---|---|---|---|---|
| V24/EIA232 | controller and instruments | n/a | n/a | 2K | 20 | n/a | 25-pin D chassis | V24 EIA232 |
| EIA422/EIA423 | controller and instruments | n/a | n/a | 100K | 1000 | n/a | 25-pin D chassis | EIA422 EIA423 |
| EIA449 | controller and instruments | n/a | n/a | 1M | 1000 | n/a | 37-pin D 9-pin D | EIA449 |
| S100 | printed circuit boards controller and instruments | 16/24 | 8/16 | 1M | 10 | 254 by 130 | 100-pin edge | IEEE696 |
| Multibus | printed circuit boards controller and instruments systems | 24 | 8/16 | 10M | 10 | 305 by 171 | 86-pin edge | IEEE796 |
| PC expansion bus | printed circuit boards controller and instruments systems | 20 | 8 | | n/a | 335 by 106 | 62-pin edge | IBM |
| STEbus | printed circuit boards instruments controller and instruments systems | 20 | 8 | 5M | 10 | Eurocard | 96-pin backplane | IEEE1000 |
| GPIB | instruments controller and instruments | 32 | 8 | 1M | 20 | n/a | 24/25-pin chassis | IEC625 IEEE488 |
| VMEbus | printed circuit boards instruments controller and instruments systems | 24/32 | 8/16/32 | 10M | 10 | Eurocard | two 96-pin backplane | IEC821 IEEE1014 |
| VXIbus | printed circuit boards | 24/32 | 8/16/32 | 1G | 10 | Eurocard | three 96-pin backplane | n/a |

Table 6.1 lists these bus systems with the levels of data communications just described, together with numbers of address and data bits, maximum data rate, maximum range and system printed circuit board sizes. The first three buses are serial with data transmitted bit by bit. The remainder are parallel, with data transmitted byte by byte.

The last three bus systems; the general-purpose interface bus, VMEbus and VXIbus, are described in following chapters. Superficial descriptions of other buses of Table 6.1 now follow.

### V24/EIA232

The V24/EIA232 interface is used regularly to transfer data between computers, or between computers and peripheral instruments. Transfer is asynchronous or synchronous, and can be a variety of operating modes: transmit only; receive only; half-duplex; full-duplex.

Users come into contact most regularly with the interface where computers or computer-based equipment transmit data over the analog telephone network using modems. Here the interface between the computer and modem is standardized worldwide by CCITT recommendation V24 (and in North America by ANSI EIA232 – which used to be called RS232 and often still is, incorrectly). In recommendation V24 the computer or equipment which generates and receives data is known as **data terminating equipment** (DTE), while the equipment which terminates the telephone line (the modem) is called **date circuit–terminating equipment** (DCE). Recommendation V24 defines the basic signal interchanges and functions between DTE and DCE; these are known as the **100 series** interchange circuits, listed in Table 6.2. Where the modem automatically calls and answers, a further recommendation (V25) defines the extra circuits required which are known as the **200 series** interchange circuits, listed in Table 6.3.

North American equivalent, EIA232, similarly defines the interface between DTE and DCE. Although the two have different designations, they are to all practical purposes interchangeable and equivalent. EIA232 interchange circuits are listed in Table 6.4. Figure 6.1 shows pin connections of the V24 25-pin D-connector.

### EIA422/EIA423

Improving on the performance of the V24/EIA232 interface, EIA422 and EIA423 provide better line matching which reduces reflections along the transmission line allowing higher data rates and line lengths to be used between DTE and DCE, while maintaining the same interchange circuits.

EIA422 specifies a balanced interface in which differential signal lines are used, terminated by a impedance as low as $50\,\Omega$. EIA423, on the other hand, specifies an unbalanced interface terminated by a $450\,\Omega$ impedance.

**Table 6.2** V24 100 series interchange circuits

| Interchange circuit | | Data | | Control | | Timing | |
|---|---|---|---|---|---|---|---|
| Number | Name | From DCE | To DCE | From DCE | To DCE | From DCE | To DCE |
| 101 | Protective ground or earth | | | | | | |
| 102 | Signal ground or common return | | | | | | |
| 103 | Transmitted data | | ● | | | | |
| 104 | Received data | ● | | | | | |
| 105 | Request to send | | | | ● | | |
| 106 | Ready for sending | | | ● | | | |
| 107 | Data set ready | | | ● | | | |
| 108/1 | Connect data set to line | | | | ● | | |
| 108/2 | Data terminal ready | | | | ● | | |
| 109 | Data channel received line signal detector | | | ● | | | |
| 110 | Signal quality detector | | | ● | | | |
| 111 | Data signalling rate selector (DTE) | | | | ● | | |
| 112 | Data signalling rate selector (DCE) | | | ● | | | |
| 113 | Transmitter signal element timing (DTE) | | | | | | ● |
| 114 | Transmitter signal element timing (DCE) | | | | | ● | |
| 115 | Receiver signal element timing (DCE) | | | | | ● | |
| 116 | Select standby | | | | ● | | |
| 117 | Standby indicator | | | ● | | | |
| 118 | Transmitted backward channel data | | ● | | | | |
| 119 | Received backward channel data | ● | | | | | |
| 120 | Transmit backward channel line signal | | | | ● | | |
| 121 | Backward channel ready | | | ● | | | |
| 122 | Backward channel received line signal detector | | | ● | | | |
| 123 | Backward channel signal quality detector | | | ● | | | |
| 124 | Select frequency groups | | | | ● | | |
| 125 | Calling indicator | | | ● | | | |
| 126 | Select transmit frequency | | | | ● | | |
| 127 | Select receive frequency | | | | ● | | |
| 128 | Receiver signal element timing (DTE) | | | | | | ● |
| 129 | Request to receive | | | | ● | | |
| 130 | Transmit backward tone | | | | ● | | |
| 131 | Received character timing | | | | | ● | |
| 132 | Return to non-data mode | | | | ● | | |
| 133 | Ready for receiving | | | | ● | | |
| 134 | Received data present | | | ● | | | |
| 191 | Transmitted voice answer | | | | ● | | |
| 192 | Received voice answer | | | ● | | | |

**Table 6.3**  V 25 200 series interchange circuits

| Interchange Circuit Number | Name | From DCE | To DCE |
|---|---|:---:|:---:|
| 201 | Signal ground | ● | ● |
| 202 | Call request | | ● |
| 203 | Data line occupied | ● | |
| 204 | Distant station connected | ● | |
| 205 | Abandon call | ● | |
| 206 | Digit signal ($2^0$) | | ● |
| 207 | Digit signal ($2^1$) | | ● |
| 208 | Digit signal ($2^2$) | | ● |
| 209 | Digit signal ($2^3$) | | ● |
| 210 | Present next digit | ● | |
| 211 | Digit present | | ● |
| 213 | Power indication | ● | |

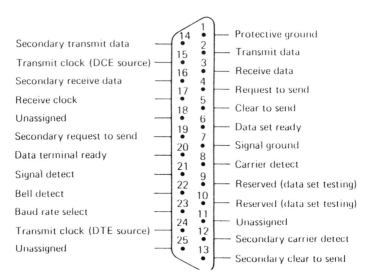

**Figure 6.1**  *Pin connections of V24 25-pin D-connector*

**Table 6.4** EIA232 interchange circuits

| Interchange circuit | | Data From To DCE DCE | | Control From To DCE DCE | | Timing From To DCE DCE | |
|---|---|---|---|---|---|---|---|
| Mnemonic | Name | | | | | | |
| AA | Protective ground | | | | | | |
| AB | Signal ground/common return | | | | | | |
| BA | Transmitted data | | ● | | | | |
| BB | Received data | ● | | | | | |
| CA | Request to send | | | | ● | | |
| CB | Clear to send | | | ● | | | |
| CC | Data set ready | | | ● | | | |
| CD | Data terminal ready | | | | ● | | |
| CE | Ring indicator | | | ● | | | |
| CF | Received line signal detector | | | ● | | | |
| CG | Signal quality detector | | | ● | | | |
| CH | Data signal rate selector (DTE) | | | | ● | | |
| CI | Data signal rate selector (DCE) | | | ● | | | |
| DA | Transmitter signal element timing (DTE) | | | | | ● | |
| DB | Transmitter signal element timing (DCE) | | | | | ● | |
| DD | Receiver signal element timing (DCE) | | | | | ● | |
| SBA | Secondary transmitted data | | ● | | | | |
| SBB | Secondary received data | ● | | | | | |
| SCA | Secondary request to send | | | | ● | | |
| SCB | Secondary clear to send | | | ● | | | |
| SCF | Secondary received line signal detector | | | ● | | | |

Specified matching terminations allow considerable performance improvements compared with the V24 interface. Data rates of around 100 Kbyte s$^{-1}$, and total distances of 1000 m are possible. Further, the interfaces allow more than one remote peripheral to be connected to the bus.

### EIA449

Improving further on V24/EIA232 interface performance, EIA449 uses a different series of interchange circuits, listed in Table 6.5. Two connectors (a 37-pin for the primary channel and a 9-pin D-connector for the secondary channel) are used, and the interface is capable of data rates up to around 1 Mbyte s$^{-1}$.

**Table 6.5**   EIA449 interchange circuits

| Interchange circuit | | Data | | Control | | Timing | |
| --- | --- | --- | --- | --- | --- | --- | --- |
| Mnemonic | Name | From DCE | To DCE | From DCE | To DCE | From DCE | To DCE |
| SC | Signal ground | | | | | | |
| SC | Send common | | | | | | |
| RC | Receive common | | | | • | | |
| IS | Terminal in service | | | • | | | |
| IC | Incoming call | | | • | | | |
| TR | Terminal ready | | | | • | | |
| DM | Data mode | | | • | | | |
| SD | Send data | | • | | | | |
| RD | Receive data | • | | | | | |
| TT | Terminal timing | | | | | | • |
| ST | Send timing | | | | | • | |
| RT | Receive timing | | | | | • | |
| RS | Request to send | | | | • | | |
| CS | Clear to send | | | • | | | |
| RR | Receiver ready | | | • | | | |
| SQ | Signal quality | | | • | | | |
| NS | New signal | | | | • | | |
| SF | Select frequency | | | | • | | |
| SR | Signalling rate selector | | | | • | | |
| SI | Signalling rate indicator | | | • | | | |
| SSD | Secondary send data | | • | | | | |
| SRD | Secondary receive data | • | | | | | |
| SRS | Secondary request to send | | | | • | | |
| SCS | Secondary clear to send | | | • | | | |
| SRR | Secondary receiver ready | | | • | | | |
| LL | Local loopback | | | | • | | |
| RL | Remote loopback | | | | • | | |
| TM | Test mode | | | • | | | |
| SS | Select standby | | | | • | | |
| SB | Standby indicator | | | • | | | |

*Primary channel* (SD through SI)

*Secondary channel* (SSD through SRR)

## S100

One of the first standardized computer buses, S100 was originally developed for 8080-based microcomputer systems. It uses a Eurocard card-frame chassis with a backplane connected to double-sided 100-pin printed circuit board edge connectors, and boards are simply slotted into the chassis as required, creating a very adaptable system. Up to 22 boards may be used in a single system. This format has been used in many microprocessor-based computer systems since.

As the name indicates, S100 systems have 100 bus lines, all designated with particular purposes within a computer system, although not all need to be used in a given system. Originally all lines were unidirectional, input and output being separate, but a revision ganged the two 8-bit data in and out lines into a single 16-bit bidirectional bus. This and other inclusions are adopted in the EIA696 standard.

The overall structure of the bus system comprises:

- 16 data lines.
- 24 address lines.
- 19 control lines.
- 8 interrupt lines.
- 8 status lines.
- 20 utility lines.
- 5 unregulated power lines.
- 4 lines reserved for future use.

Table 6.6 lists pin assignments of the S100, while Figure 6.2 illustrates pin numbering used for the printed circuit board connector.

**Table 6.6**  S100 edge connector pin assignments

| Pin no. | Abbreviation | Signal/function |
|---------|--------------|-----------------|
| 1 | +8 V | Unregulated supply rail |
| 2 | +18 V | Unregulated supply rail |
| 3 | XRDY | Ready input to bus master |
| 4 | VI0 | Vectored interrupt line 0 |
| 5 | VI1 | Vectored interrupt line 1 |
| 6 | VI2 | Vectored interrupt line 2 |
| 7 | VI3 | Vectored interrupt line 3 |
| 8 | VI4 | Vectored interrupt line 4 |
| 9 | VI5 | Vectored interrupt line 5 |
| 10 | VI6 | Vectored interrupt line 6 |
| 11 | VI7 | Vectored interrupt line 7 |
| 12 | NMI | Non-maskable interrupt |

**Table 6.6**  (*cont.*)

| Pin no. | Abbreviation | Signal/function |
|---|---|---|
| 13 | PWRFAIL | Power fail signal (pulled low when a power failure is detected) |
| 14 | DMA3 | DMA request line with highest priority |
| 15 | A18 | Extended address bus line 18 |
| 16 | A16 | Extended address bus line 16 |
| 17 | A17 | Extended address bus line 17 |
| 18 | SDSB | Status disable (tri-states all status lines) |
| 19 | CDSB | Command disable (tri-states all control input lines) |
| 20 | GND | Common 0 V line |
| 21 | NDEF | Undefined |
| 22 | ADSB | Address disable (tri-states all address lines) |
| 23 | DODSB | Data out disable (tri-states all data output lines) |
| 24 | 0 | Bus clock |
| 25 | $\rho$STVAL | Status valid strobe (indicates that status information is true) |
| 26 | $\rho$HLDA | Hold acknowledge (signal from the current bus master which indicates that control will pass to the device seeking bus control on the next bus cycle) |
| 27 | RFU | Reserved for future use |
| 28 | RFU | Reserved for future use |
| 29 | A5 | Address line 5 |
| 30 | A4 | Address line 4 |
| 31 | A3 | Address line 3 |
| 32 | A15 | Address line 15 |
| 33 | A12 | Address line 12 |
| 34 | A9 | Address line 9 |
| 35 | DO1/Data 1 | Data out line 1/bidirectional data line 1 |
| 36 | DO0/Data 0 | Data out line 0/bidirectional data line 0 |
| 37 | A10 | Address line 10 |
| 38 | DO4/Data 4 | Data out line 4/bidirectional data line 4 |
| 39 | DO5/Data 5 | Data out line 5/bidirectional data line 5 |
| 40 | DO6/Data 6 | Data out line 6/bidirectional data line 6 |
| 41 | DI2/Data 10 | Data in line 2/bidirectional data line 10 |
| 42 | DI3/Data 11 | Data in line 3/bidirectional data line 11 |
| 43 | DI7/Data 15 | Data in line 7/bidirectional data line 15 |
| 44 | sM1 | M1 cycle (indicates that the current machine cycle is in an operation code fetch) |
| 45 | sOUT | Output (indicates that data is being transferred to an output device) |
| 46 | sINP | Input (indicates that data is being fetched from an input device) |
| 47 | sMEMR | Memory read (indicates that the bus master is fetching data from memory) |

**Table 6.6** *(cont.)*

| Pin no. | Abbreviation | Signal/function |
|---------|--------------|-----------------|
| 48 | sHLTA | Halt acknowledge (indicates that the bus master is executing an HLT instruction) |
| 49 | CLOCK | 2 MHz clock |
| 50 | GND | Common 0 V |
| 51 | +8 V | Unregulated supply rail |
| 52 | −16 V | Unregulated supply rail |
| 53 | GND | Common 0 V |
| 54 | SLAVE CLR | Slave clear (resets all bus slaves) |
| 55 | DMA0 | DMA request line (lowest priority) |
| 56 | DMA1 | DMA request line |
| 57 | DMA2 | DMA request line |
| 58 | sXTRQ | 16-bit data request (requests slaves to assert SIXTN) |
| 59 | A19 | Address line 19 |
| 60 | SIXTN | 16-bit data acknowledge (slave response to sXTRQ) |
| 61 | A20 | Extended address bus line 20 |
| 62 | A21 | Extended address bus line 21 |
| 63 | A22 | Extended address bus line 22 |
| 64 | A23 | Extended address bus line 23 |
| 65 | NDEF | Not defined |
| 66 | NDEF | Not defined |
| 67 | PHANTOM | Phantom (disables normal slaves and enables phantom slaves which share addresses with the normal set) |
| 68 | MWRT | Memory write |
| 69 | RFU | Reserved for future use |
| 70 | GND | Common 0 V |
| 71 | RFU | Reserved for future use |
| 72 | RDY | Ready input to bus master |
| 73 | INT | Interrupt request |
| 74 | HOLD | Hold request (request from device wishing to have control of the bus) |
| 75 | RESET | Reset (resets bus master devices) |
| 76 | $\rho$SYNC | Synchronising signal which indicates the first bus state of a bus cycle |
| 77 | $\rho$WR | Write (indicates that the bus master has placed valid data on the DO bus/data bus) |
| 78 | $\rho$DBIN | Data bus in (indicates that the bus master is requesting data on the DI bus/data bus) |
| 79 | A0 | Address line 0 |
| 80 | A1 | Address line 1 |
| 81 | A2 | Address line 2 |
| 82 | A6 | Address line 6 |
| 83 | A7 | Address line 7 |

**Table 6.6** (*cont.*)

| Pin no. | Abbreviation | Signal/function |
|---|---|---|
| 84 | A8 | Address line 8 |
| 85 | A13 | Address line 13 |
| 86 | A14 | Address line 14 |
| 87 | A11 | Address line 11 |
| 88 | DO2/DATA 2 | Data out line 2/bidirectional data line 2 |
| 89 | DO3/DATA 3 | Data out lines 3/bidirectional data line 3 |
| 90 | DO7/DATA 7 | Data out line 7/bidirectional data line 7 |
| 91 | D14/DATA 12 | Data in line 4/bidirectional data line 12 |
| 92 | DI5/DATA 13 | Data in line 5/bidirectional data line 13 |
| 93 | DI6/DATA 14 | Data in line 6/bidirectional data line 14 |
| 94 | DI1/DATA 9 | Data in line 1/bidirectional data line 9 |
| 95 | DI0/DATRA 8 | Data in line 0/bidirectional data line 8 |
| 96 | sINTA | Interrupt acknowledge |
| 97 | sWO | Write output (used to gate data from the bus master to a slave) |
| 98 | ERROR | Error (indicates that an error has occurred during the current bus cycle) |
| 99 | POC | Power on clear (clears all devices attached to the bus when power is first applied) |
| 100 | GND | Common 0 V |

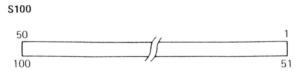

**Figure 6.2** *Pin numbering in the S100 printed circuit board edge connector*

## Multibus

Similar to the S100 bus, designed by Intel for its 8086 range of microprocessors, the Multibus is an 86-line bus based on printed circuit board edge connectors. Two edge connectors are used on each Eurocard-based board: a P1 connector, carrying all 86 lines of the bus; a P2 connector carrying 60 lines for custom features. The overall bus structure comprises:

- 16 data lines.
- 24 address lines.

- 8 interrupt lines.
- 17 control lines.
- 5 power lines.
- 2 lines reserved for future use.

Some power lines are found on more than one pin of the edge connector.

Multibus is defined in the standard IEEE796 and is quite a popular industrial computer-based bus. Table 6.7 lists pin assignments of Multibus, while Figure 6.3 illustrates pin numbering of the P1 edge connector.

## PC expansion bus

IBM PC is an extremely popular computer, and it is no surprise that it has been used as the central control of automatic test equipment systems. Generally, this is accomplished using the PC expansion bus, which features a 62-pin printed circuit board edge connector.

The overall bus structure comprises:

- 8 data lines.
- 20 address lines.
- 8 interrupt lines.
- 18 control lines.
- 5 power lines.
- 1 line reserved for future use.

Table 6.8 lists pin assignments of the PC expansion bus, while Figure 6.4 illustrates pin numbering of the edge connector.

**Figure 6.3** *Pin numbering in the Multibus printed circuit board edge connector*

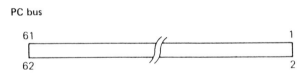

**Figure 6.4** *Pin numbering of the IBM PC expansion bus printed circuit board edge connector*

**Table 6.7**    Multibus edge connector pin assignments

*Component side*

| Pin no. | Signal group | Abbreviation | Signal/function |
|---|---|---|---|
| 1 | Supply | GND | Ground/common 0 V |
| 3 | rails | +5 V | +5 V DC supply rail |
| 5 | | +5 V | +5 V DC supply rail |
| 7 | | +12 V | +12 V DC supply rail |
| 9 | | −5 V | −5 V DC supply rail |
| 11 | | GND | Ground/common 0 V |
| 13 | Bus | BCLK | Bus clock |
| 15 | control | BPRN | Bus priority input |
| 17 | | BUSY | Bus busy |
| 19 | | MRDC | Memory read command |
| 21 | | IORC | I/O read command |
| 23 | | XACK | Transfer acknowledge |
| 25 | | | Reserved |
| 27 | | BHEN | Byte high enable |
| 29 | | CBRQ | Common bus request |
| 31 | | CCLK | Constant clock |
| 33 | | INTA | Interrupt acknowledge |
| 35 | Interrupt | INT6 | Parallel interrupt requests |
| 37 | | INT4 | |
| 39 | | INT2 | |
| 41 | | INT0 | |
| 43 | Address | ADRE | Address line |
| 45 | bus | ADRC | |
| 47 | | ADRA | |
| 49 | | ADR8 | |
| 51 | | ADR6 | |
| 53 | | ADR4 | |
| 55 | | ADR2 | |
| 57 | | ADR0 | |
| 59 | Data | DATE | Data lines |
| 61 | bus | DATC | |
| 63 | | DATA | |
| 65 | | DAT8 | |
| 67 | | DAT6 | |
| 69 | | DAT4 | |
| 71 | | DAT2 | |
| 73 | | DAT0 | |
| 75 | Supply | GND | Ground/common 0 V |
| 77 | rails | | Reserved |
| 79 | | −12 V | −12 V DC supply rail |
| 81 | | +5 V | +5 V DC supply rail |
| 83 | | +5 V | +5 V DC supply rail |
| 85 | | GND | Ground/common 0 V |

**Table 6.7** *(cont.)*

| Track side Pin no. | Signal group | Abbreviation | Signal/function |
|---|---|---|---|
| 2 | Supply | GND | Ground/common 0 V |
| 3 | rails | +5 V | +5 V DC supply rail |
| 6 | | +5 V | +5 V DC supply rail |
| 8 | | +12 V | +12 V DC supply rail |
| 10 | | −5 V | −5 V DC supply rail |
| 12 | | GND | Ground/common 0 V |
| 14 | Bus | INIT | Initialize |
| 16 | control | BPRO | Bus priority output |
| 18 | | BREQ | Bus request |
| 20 | | MWTC | Memory write command |
| 22 | | IOWC | I/O write command |
| 24 | | INH1 | Inhibit 1 (diable RAM) |
| 26 | | INH2 | Inhibit 2 (diable ROM) |
| 28 | Address | AD10 | Address lines |
| 30 | bus | AD11 | |
| 32 | | AD12 | |
| 34 | | AD13 | |
| 36 | Interrupt | INT7 | Parallel interrupt requests |
| 38 | | INT5 | |
| 40 | | INT3 | |
| 42 | | INT1 | |
| 44 | Address | ADRF | Address lines |
| 46 | Bus | ADRD | |
| 48 | | ADRB | |
| 50 | | ADR9 | |
| 52 | | ADR7 | |
| 54 | | ADR5 | |
| 56 | | ADR3 | |
| 58 | | ADR1 | |
| 60 | Data | DATF | Data lines |
| 62 | bus | DATD | |
| 64 | | DATB | |
| 66 | | DAT9 | |
| 68 | | DAT7 | |
| 70 | | DAT5 | |
| 72 | | DAT3 | |
| 74 | | DAT1 | |
| 76 | Supply | GND | Ground/common 0 V |
| 78 | rails | | Reserved |
| 80 | | −12 V | −12 V DC supply rail |
| 82 | | +5 V | +5 V DC supply rail |
| 84 | | +5 V | +5 V DC supply rail |
| 86 | | GND | Ground-common 0 V |

**Table 6.8**   IBM PC expansion bus edge connector pin assignments

| Pin no. | Abbreviation | Signal/function |
|---------|--------------|-----------------|
| 1 | GND | Ground/common 0 V |
| 2 | CHCK | Channel check output (when low this indicates that some form of error has occurred) |
| 3 | RESET | Reset (when high this line resets all expansion cards) |
| 4 | D7 | Data line 7 |
| 5 | +5 V | +5 V DC supply rail |
| 6 | D6 | Data line 6 |
| 7 | IRQ2 | Interrupt request input 2 |
| 8 | D5 | Data line 5 |
| 9 | −5 V | −5 V DC supply rail |
| 10 | D4 | Data line 4 |
| 11 | DRQ2 | DMA request input 2 |
| 12 | D3 | Data line 3 |
| 13 | −12 V | −12 V DC supply rail |
| 14 | D2 | Data line 2 |
| 15 | | Reserved |
| 16 | D1 | Data line 1 |
| 17 | +12 V | +12 V DC supply rail |
| 18 | D0 | Data line 0 |
| 19 | GND | Ground/common 0 V |
| 20 | BCRDY | Ready input (normally high, pulled low by a slow memory or I/O device to signal that it is not ready for data transfer to take place) |
| 21 | IMW | Memory write output |
| 22 | AEN | Address enable output |
| 23 | IMR | Memory read output |
| 24 | A19 | Address line 19 |
| 25 | IIOW | I/O write output |
| 26 | A18 | Address line 18 |
| 27 | IIOR | I/O read output |
| 28 | A17 | Address line 17 |
| 29 | DACK3 | DMA acknowledge output 3 (see notes) |
| 30 | A16 | Address line 16 |
| 31 | DRQ3 | DMA request input 3 |
| 32 | A15 | Address line 15 |
| 33 | DACK1 | DMA acknowledge output 1 (see notes) |
| 34 | A14 | Address line 14 |
| 35 | DRQ1 | DMA request input 1 |
| 36 | A13 | Address line 13 |
| 37 | DACK0 | DMA acknowledge output 0 (see notes) |
| 38 | A12 | Address line 12 |
| 39 | XCLK4 | 4 MHz clock (CPU clock divided by two 200 ns period, 50% duty cycle) |

**Table 6.8** *(cont.)*

| Pin no. | Abbreviation | Signal/function |
|---------|--------------|-----------------|
| 40 | A11 | Address line 11 |
| 41 | IRQ7 | Interrupt request line 7 (see notes) |
| 42 | A10 | Address line 10 |
| 43 | IRQ6 | Interrupt request line 6 (see notes) |
| 44 | A9 | Address line 9 |
| 45 | IRQ5 | Interrupt request line 5 |
| 46 | A8 | Address line 8 |
| 47 | IRQ4 | Interrupt request line 4 (see notes) |
| 48 | A7 | Address line 7 |
| 49 | IRQ3 | Interrupt request line 3 |
| 50 | A6 | Address line 6 |
| 51 | DACK2 | DMA acknowledge 2 |
| 52 | A5 | Address line 5 |
| 53 | TC | Terminal count output (pulsed high to indicate that the terminal count for a DMA transfer has been reached) |
| 54 | A4 | Address line 4 |
| 55 | ALE | Address latch enable output |
| 56 | A3 | Address line 3 |
| 57 | +5 V | +5 V DC supply rail |
| 58 | A2 | Address line 2 |
| 59 | 14 MHz | 14.31818 MHz clock (fast clock with 70 ns period, 50% duty cycle) |
| 60 | A1 | Address line 1 |
| 61 | GND | Ground/common OV |
| 62 | IA0 | Address line 0 |

*Notes:* (a) Signal direction is quoted relative to the motherboard
(b) IRQ4 is generated by the motherboard serial interface
IRQ6 is generated by the motherboard disk interface
IRQ7 is generated by the motherboard parallel interface
(c) DACK0 is used to refresh dynamic memory, while DACK1 to DACK3 are used to acknowledge DMA requests.

## STEbus

STEbus is the successor to an earlier bus; STDbus, upgraded and redesigned to allow backplaned construction using standard Eurocard-sized modules. Indeed, the name STEbus is merely a contraction of STDbus and Eurocard. It was defined in 1982 when the Institute of Electrical and Electronic Engineers set up a working group to develop it, eventually specifying it in 1984. It became standardized in 1987 by IEEE1000.

Original STDbus (produced in 1978, and standardized as IEEE961) is specifically for 8-bit microprocessors, so is fairly limited. On the other hand, while still having only 8 data lines, STEbus features a 20-bit address bus and a maximum data rate of around 5 Mbyte s$^{-1}$. It is more of an industrial control bus system than one for instrumentation, but can be and is used for automatic test equipment systems. In terms of popularity, STEbus is known to be the fastest-growing 8-bit bus, particularly in the UK, although it is becoming more popular elsewhere as its advantages over similar bus systems are accepted.

As a bus system STEbus is flexible; it is microprocessor-independent yet precisely defines bus signals, levels and protocols. Further, its use of Eurocard-based plug-in boards means STEbus systems may be modular in concept. In this respect, it may be viewed as a scaled-down version of 16-bit or 32-bit standard bus systems such as VMEbus or VXIbus.

The overall bus structure includes:

- 8 data lines.
- 20 address lines.
- 20 control lines.
- 4 power supply lines.
- 1 clock line.

Connectors used for the modular construction are standard connectors (defined in IEC603-2) with three rows of pins, 32 pins in each row. This is the first bus system considered which uses such connectors. The connector is shown in Figure 6.5, with pin assignments. Not the middle row (row B) of the connector is not defined in IEEE1000 and is for customized usage. These connectors are the same as those used on VMEbus and VXIbus, and it does not take much consideration to realize STEbus and VMEbus devices may be combined in the same chassis (this probably adds to the earlier view of STEbus being a scaled-down version of VMEbus). For such purposes a VMEbus-to-STEbus coupler board is used, which plugs into a VMEbus backplane P2 connector to link the two buses, as shown in Figure 6.6.

IEEE1000 is quite specific on four types of device which may be connected to the bus: system controller; arbiter; master; slave.

### System controller

Only one system controller may exist in any system. It provides three important signals:

- SYSCLK, a general-purpose clock.
- SYSRST★, a power-on reset.
- TRFERR★, a transfer error.

These signals may be provided as part of an STEbus master device, however.

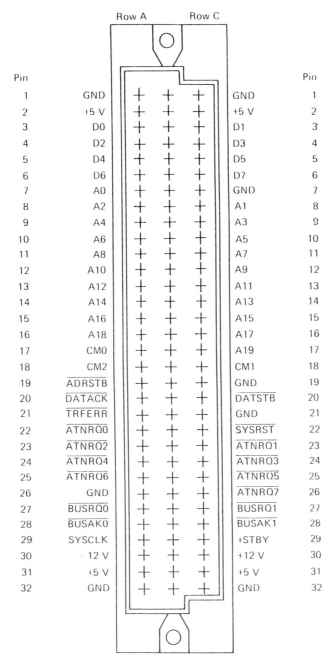

| Pin | Row A | | | | Row C | Pin |
|---|---|---|---|---|---|---|
| 1 | GND | + | + | + | GND | 1 |
| 2 | +5 V | + | + | + | +5 V | 2 |
| 3 | D0 | + | + | + | D1 | 3 |
| 4 | D2 | + | + | + | D3 | 4 |
| 5 | D4 | + | + | + | D5 | 5 |
| 6 | D6 | + | + | + | D7 | 6 |
| 7 | A0 | + | + | + | GND | 7 |
| 8 | A2 | + | + | + | A1 | 8 |
| 9 | A4 | + | + | + | A3 | 9 |
| 10 | A6 | + | + | + | A5 | 10 |
| 11 | A8 | + | + | + | A7 | 11 |
| 12 | A10 | + | + | + | A9 | 12 |
| 13 | A12 | + | + | + | A11 | 13 |
| 14 | A14 | + | + | + | A13 | 14 |
| 15 | A16 | + | + | + | A15 | 15 |
| 16 | A18 | + | + | + | A17 | 16 |
| 17 | CM0 | + | + | + | A19 | 17 |
| 18 | CM2 | + | + | + | CM1 | 18 |
| 19 | $\overline{ADRSTB}$ | + | + | + | GND | 19 |
| 20 | $\overline{DATACK}$ | + | + | + | $\overline{DATSTB}$ | 20 |
| 21 | $\overline{TRFERR}$ | + | + | + | GND | 21 |
| 22 | $\overline{ATNRQ0}$ | + | + | + | $\overline{SYSRST}$ | 22 |
| 23 | $\overline{ATNRQ2}$ | + | + | + | $\overline{ATNRQ1}$ | 23 |
| 24 | $\overline{ATNRQ4}$ | + | + | + | $\overline{ATNRQ3}$ | 24 |
| 25 | $\overline{ATNRQ6}$ | + | + | + | $\overline{ATNRQ5}$ | 25 |
| 26 | GND | + | + | + | $\overline{ATNRQ7}$ | 26 |
| 27 | $\overline{BUSRQ0}$ | + | + | + | $\overline{BUSRQ1}$ | 27 |
| 28 | $\overline{BUSAK0}$ | + | + | + | $\overline{BUSAK1}$ | 28 |
| 29 | SYSCLK | + | + | + | +STBY | 29 |
| 30 | – 12 V | + | + | + | +12 V | 30 |
| 31 | +5 V | + | + | + | +5 V | 31 |
| 32 | GND | + | + | + | GND | 32 |

**Figure 6.5** *STEbus Eurocard-style connector, with pin numbering and assignments*

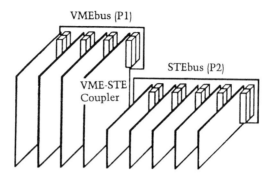

**Figure 6.6**   *Using a VMEbus-to-STEbus coupler, connecting between an STEbus backplane and the P2 connector of a VMEbus backplane*

### Arbiter

Only one arbiter may exist in a system. It provides bus grant signals in response to requests for use of the bus by masters. These signals, like those of a system controller, may be provided as part of an STEbus master device.

### Master

An STEbus master device is capable of controlling data transfer over the STEbus. All masters must request and receive permission to control the bus from the system arbiter, before transmission is permitted. The time that masters are allowed to control bus operation depends on which of two operational modes masters are allotted:

- **Release-when-done**, where a master device maintains bus control until all its data transmissions are complete.
- **Release-on-request**, where a master device maintains bus control until another master requests control.

### Slaves

As its name suggests, an STEbus slave merely acts upon the instruction of other devices on the bus.

STEbus backplanes are fairly complex, carrying all connectors, all signals between boards, and power to all boards. A maximum of 21 boards may be plugged into the STEbus chassis.

# 7 General-purpose interface bus

Arguably the most influential of all instrument buses is the general-purpose interface bus (GPIB). Although it is currently in the process of being superseded by a more powerful and more appropriate bus system – in the form of VXIbus – the general-purpose interface bus has been, for some 25 years, an almost natural choice when automatic test equipment systems have been proposed. Further, even though VXIbus is without doubt a much superior bus system, the general-purpose interface bus is still being used and indeed will be used in automatic test equipment systems for many years to come.

General-purpose interface bus owes its existence to Hewlett Packard, the company which developed it in 1965. Originally known as **Hewlett Packard interface bus** (HP-IB) it was used first in Hewlett Packard's own programmable measuring instruments. It quickly became a most popular method of interfacing instruments and many manufacturers began to incorporate bus interfaces into their own instruments, giving users the option to mix and match different manufacturers' equipments.

It was adopted by the American Institute of Electrical and Electronic Engineers in the IEEE488 standard in 1975, and the American National Standards Institute incorporated it as the ANSI MC1.1 standard. Later, as standard IEC625, the International Electrotechnical Commission adopted it (BS6146 in the UK).

Worldwide standards such as these mean the bus is accepted internationally with great credibility. Most automatic test equipment manufacturers now produce instruments featuring a bus interface, which means individual instruments are effectively programmable at a distance across the bus and so may be used as part of an automatic test equipment system, controlled by a computer.

This fact – users can build automatic test equipment systems out of individual parts, to their own specifications – revolutionized automatic test equipment. Such **rack and stack** systems may be designed, built,

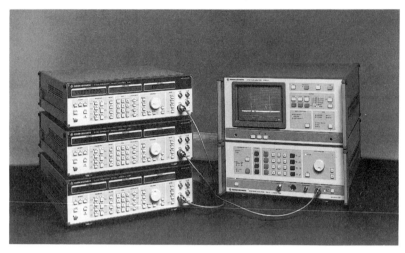

**Photo 7.1** *Test equipment, based around the general-purpose interface bus, combined in a rack and stack automatic test equipment system (Rohde & Schwarz)*

**Photo 7.2** *Rack and stack automatic test equipment, based around the Fluke 9020A troubleshooter (Philips)*

programmed and maintained by the user. Systems are under user control, not manufacturer control.

There are disadvantages, on the other hand. First, size — a system comprising more than just a few instruments takes considerable work-bench area. Second, data speed – system data speed is generally much less than the general-purpose interface bus quoted maximum of 1 Mbyte s$^{-1}$; not really good enough to build a system which needs to perform many tests on many nodes rapidly. Third, not everyone has or desires the necessary expertise required to design, build, program and maintain such a system; often buying in a purpose-built turnkey automatic test equipment system along with a maintenance contract is a preferred option. Finally, price – a large system comprising many expensive instruments may cost an arm and a leg!

These disadvantages prompted VXIbus, which looks set to overtake the general-purpose interface bus in overall popularity for use in automatic test equipment systems. Nevertheless, where certain kinds of automatic test equipment system are concerned, general-purpose interface bus is still most effective; and will continue to be so for many years yet.

## Bus structure

General-purpose interface bus standards define all hardware characteristics (that is, cable, connectors, voltage and current values of signals, purposes of the signals, and timing relationships between the signals) of the interface, but leave remaining software characteristics to the user. In other words connecting automatic test equipment system components together is simply a matter of plugging them in, but controlling system operation relies on the user buying or writing software required to perform a measurement task. Generally speaking, instruments with different data rate capabilities are adequately interfaced by the bus, so as long as an instrument has a general-purpose interface bus interface it can be used with considerable certainty.

Design is of a structured interface, with many outlets joined by a single cable. Figure 7.1 illustrates general-purpose interface bus structure. Till now, only programmable test instruments have been discussed as connect-ing to a general-purpose interface bus. However, the bus is, in reality, a type of computer bus, passing data and commands much like a purpose-designed computer bus does. Consequently, programmable test instru-ments form not the only type of component which may be connected. System components of any type are usually referred to, for this reason, as **devices** by relevant standards and literature.

All devices are connected in parallel to the cable of the bus, and all devices have access to all lines making up the bus. Connection between a

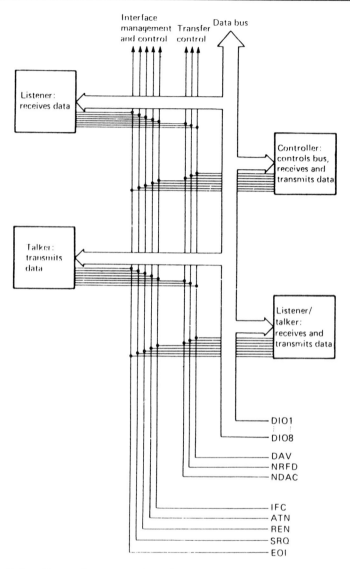

**Figure 7.1**　*General-purpose interface bus structure. One controller manages all communications between devices connected to the bus*

device and the general-purpose interface bus is by a single, standard type of connector. This is a simple locking plug-and-socket arrangement governed by international standards. Connector pin assignments, electrical and mechanical details are detailed later. Inside devices a microprocessor-based circuit deals with electrical interfacing. Electrical device interfacing is discussed later.

Any device on the bus may (depending on its capabilities) send or receive data to or from any other device; also, a sender of data may send data, simultaneously, to more than one receiver. Devices connected to a general-purpose interface bus system fall into one of the following categories:

- A **listener** – can only receive data.
- A **talker** – can only send data.
- A **listener/talker** – can be switched between listening and talking.
- A **controller** – a computer; to determine which devices talk and which listen during data transfers. The controller also has the task of sending special commands called **interface messages** to devices on the bus. Only one controller is necessary on a bus.

These are, however, fairly arbitrary categories and borders between them may be blurred somewhat by technology.

## Listeners
Devices which only receive data from the general-purpose interface bus are few and far between. A printer is an obvious example, indicating lamps and robotic controllers of various descriptions are others. Signal sources and programmable power supplies are two other examples.

## Talkers
Talkers are typically fairly straightforward measuring instruments, such as digital voltmeters, frequency counters or oscilloscopes. These are probably set up manually by the user in the first instance, to measure and indicate the measurand chosen to be studied, prior to setting up the complete system. Measured results from talkers are placed on the bus for use by listeners.

## Listener/talkers
It is this category of device which is most arbitrary. Generally these are programmable instruments which are directly programmed by the controlling computer. Thus, digital voltmeters, frequency counters or oscilloscopes (note the same examples as in the talker category) which are set up under programmed control through the general-purpose interface bus are classed as listener/talkers. In effect, these devices *listen* while the computer issues commands to set them up to measure specific measurands, then *talk* while performing the measurements – sending back readings taken to the computer.

## Controllers
Controllers monitor and control data flow across the general-purpose interface bus and allocate specific times when devices are to communicate with each other. Bus management commands are also specifically performed by controllers. Typically a personal computer is used as a system

controller, though purpose-built controllers are available, as are combined test instruments and controllers. In certain cases, a controller is not an essential part of a system. If, say, devices can be manually set to listen-only, while a single talker talks, a simple system can be constructed without need of a controller.

## Devices: functions and messages

Messages passed on the general-purpose interface bus represent quantities of information. What information and what is implied by any particular message depends on its purpose. To understand how messages are used on a system, Figure 7.2 illustrates a link between any two devices on the general-purpose interface bus. This shows two areas of interest, usually thought of as **functional** areas within devices.

- **Device functions** – functions which depend totally on the devices connected by the interface. Consequently, they are often defined when required, rather than being standardized. Standards do exist, on the other hand, which are usually referred to as recommendations rather than strict rules.
- **Interface functions** – functions which allow a device to receive, process or send a message. There are ten interface functions (see later).

This use of conceptual functions to illustrate device operation is to say the very least, ambiguous. Standards which define the general-purpose interface bus, however, use this ambiguous method, so we are stuck with it! It remains the major cause of misunderstanding when considering general-purpose interface bus operation, as even standards do not explain the concept adequately. It is important to remember:

*functional areas are conceptual and not necessarily split, mechanically or electrically, within devices.*

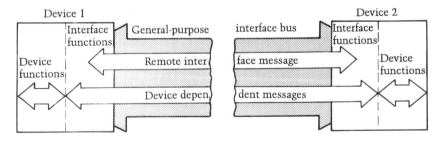

**Figure 7.2**   *A link between two devices connected by the general-purpose interface bus. Device functions and interface functions are important functional areas*

Messages pass in a number of ways on the general-purpose interface bus. There are two main types of messages:

- **Remote messages** – passed between interface functions of different devices.
- **Local messages** – passed between device functions and interface functions of single devices. These depend on the device and its application, so are defined by the manufacturer.

In turn, there are two types of remote messages:

- **Interface messages** – passed purely between interface functions of different devices.
- **Device-dependent messages** – passed between device functions of different devices, via interface functions. These are effectively local messages transmitted between remote devices and, like local messages, depend on the device and its application, so are defined by the manufacturer.

Figure 7.3 illustrates functional areas and message paths in a device, along with required linkages between interface functions, message coding logic, drivers and receivers. Message paths are numbered, and are listed together with explanations in Table 7.1. Interface functions are shown in abbreviated form, and are listed with explanations and message paths in Table 7.2.

## Message coding

Messages are coded for transmission over general-purpose interface bus signal lines. A message transmitted over a single signal line is known as a **uniline message**, while a message transmitted over a group of signal lines (as a data byte over a data bus) is known as a **multiline message**.

As multiline messages are of either interface message or device-dependent message forms, some way of differentiating between the two must be available. This is done by using the logic state of one or more separate signal lines (see later) as a uniline message to inform listeners which form of message is to be transmitted.

## Interface

The general-purpose interface bus consists of 16 signal lines, divided into three groups, comprising:

- 8 data lines for transfer of measurement data, commands and addresses.
- 3 lines for data transfer control.
- 5 lines for interface management and control.

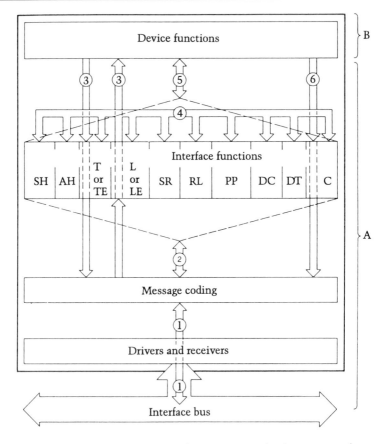

**Figure 7.3** *Functional areas and message paths in a general-purpose interface bus device*

**Table 7.1**  Key to functional areas and message paths of Figure 7.3

| Number | Message path |
| --- | --- |
| 1 | interface bus signal lines |
| 2 | interface messages to and from interface functions |
| 3 | device-dependent messages to and from device functions |
| 4 | state linkages between interface functions |
| 5 | local messages between device functions and interface functions |
| 6 | interface messages sent by device functions in a controller |

**Table 7.2** Interface functions of a general-purpose interface bus device

| Interface symbol | Interface function | Message paths |
|---|---|---|
| SH | source handshake | 1,2,4,5 |
| AH | acceptor handshake | 1,2,4,5 |
| T or TE | talker or extended talker | 1,2,3,4,5 |
| L or LE | listener or extended listener | 1,2,3,4,5 |
| SR | service request | 1,2,4,5 |
| RL | remote local | 1,2,4,5 |
| PP | parallel poll | 1,2,4,5 |
| DC | device clear | 1,2,4,5 |
| DT | device trigger | 1,2,4,5 |
| C | controller | 1,2,4,5,6 |

Data lines DIO1 to DIO8 form a bidirectional data bus. Data is transferred byte by byte (that is, byte-serial); while each byte's eight bits are transferred in parallel (bit-parallel). Bit-parallel, byte-serial transmission is a common enough digital communications method. Generally, ASCII encoded information is transmitted across the 8 data line bus (seven bits for ASCII code, one bit as a parity check digit), allowing a standard range of upper and lower alphabetic characters, numeric characters, signs, punctuation marks and control characters numbering 128 in all. Bytes are exchanged between an enabled talker and an enabled listener in a handshaking sequence, described later.

These data lines carry device-dependent data directly to or from devices, as well as specific interface messages for device functions. Which type of information is carried at any time is indicated to all devices by the state of the *attention* signal (see later).

The three data transfer control lines are now described.

## Data valid (DAV)
This signal indicates valid data from a talker is present on data lines, ready to be sent to listeners.

## Not ready for data (NRFD)
This signal indicates not all listeners are ready to receive data. Data transfer is only allowed to take place if all listeners indicate readiness to accept data, thus if any listener asserts the *not ready for data* signal the talker is inhibited from asserting a *data valid* signal.

## Not data accepted (NDAC)
This signal indicates not all listeners have received the data. The signal only

becomes unasserted when all listeners have accepted the data byte currently on data lines.

The five interface management and control lines are now described.

### Interface clear (IFC)

This signal can be asserted by the controller alone and is used to set interfaces of all peripheral units to a predefined condition. This is useful, say, immediately after switch-on.

### Attention (ATN)

This signal indicates the nature of information on data lines. A logic 0 signal indicates presence of device dependent message data: a logic 1 indicates presence of interface messages in the form of addresses or commands.

### Remote enable (REN)

This signal instructs all devices connected to the bus to be prepared for remote control operation. All device panel controls are blocked as soon as they are addressed as a listener.

### Service request (SRQ)

A signal on this line allows any device to interrupt the controller, thereby demanding attention.

### End or identity (EOI)

Alone, this signal indicates end of a multiple byte transfer sequence. When *end or identify* is used in conjunction with *attention*, the signals force the controller to execute a parallel polling sequence – see later in this chapter – which identifies the device requesting service.

## Bus operation

Command of the general-purpose interface bus is assumed by the system controller merely by asserting the *attention* (ATN) line, taking it to logic 1. All devices connected to the bus are then in listening mode, ready to act on following commands issued by the controller.

Once a controller is in command mode it uses the 8-bit data bus to transmit multiline interface messages as a 7-bit code, with an extra bit for parity check, similar to ASCII code. These interface messages are either **commands** or **addresses**. Specific commands available in command mode, which each other device on the general-purpose interface bus must react to, are therefore represented on the data bus as a single 8-bit byte, and are listed in 7-bit forms together with mnemonic names in Table 7.3.

**Table 7.3** Mnemonics of
general-purpose interface bus
commands and 7-bit forms

| Code | Mnemonic |
| --- | --- |
| 0000001 | GTL |
| 0000100 | SDC |
| 0000101 | PPC |
| 0001000 | GET |
| 0001001 | TCT |
| 0010001 | LLO |
| 0010100 | DCL |
| 0010101 | PPU |
| 0011000 | SPE |
| 0011001 | SPD |
| 0111111 | UNL |
| 1011111 | UNT |

## Commands

Commands are in two basic types; **universal commands** which are issued to all devices and, **addressed commands** which are issued to specific devices. Addresses of devices to which addressed commands are issued are transmitted with commands over the data bus.

Command mnemonics, together with mnemonic meanings, equivalent ASCII characters, descriptions and command types are listed in Table 7.4.

### Unaddressed commands
Unaddressed commands are a special type of universal command. There are two: *unlisten* (UNL) and *untalk* (UNT). Their effect is to cancel a previously addressed system condition. Thus, if there are listeners on the bus, addressed by the controller earlier, the *unlisten* command cancels the listening condition. Similarly, the *untalk* command cancels any talkers on the bus.

### Addressing
Devices are addressed as listeners or talkers with unique 7-bit codewords. These are listed in Table 7.5, with main difference between talker and listener addresses being state of bit 7; when bit 7 is logic 0 listeners are addressed, when logic 1 talkers are addressed.

### Secondary commands and addressing
There are currently two commands which must be followed by further information, directly related and giving data which devices need to be able

**Table 7.4**  General-purpose interface bus commands, mnemonics, meanings, equivalent ASCII characters and command types

| Command mnemonic | Mnemonic meaning | Equivalent ASCII character | Description | Command type |
|---|---|---|---|---|
| GTL | go to local | SOH | returns instruments to local control | addressed |
| SDC | selective device clear | EOT | returns instruments to predetermined states | addressed |
| PPC | parallel poll configure | ENQ | permits instruments to use DIO lines to respond to a parallel poll | addressed |
| GET | group execute trigger | BS | initiates simultaneous pre-programmed actions | addressed |
| TCT | take control | HT | passes control to another controller | addressed |
| LLO | local lockout | DC1 | disables instrument front panels | universal |
| DCL | device clear | DC4 | returns instruments to predetermined states | universal |
| PPU | parallel poll unconfigure | NAK | sets instruments capable of parallel poll to a predetermined condition | universal |
| SPE | serial poll enable | CAN | enables serial poll mode | universal |
| SPD | serial poll disable | EM | disables serial poll mode | universal |
| UNL | un-listen | | clears bus of listeners | |
| UNT | un-talk | | clears bus of talkers | |

**Table 7.5**  Talk and listen addresses on a general-purpose interface bus, given as 8-bit codes and equivalent ASCII characters

| Code | Equivalent ASCII character (listener) | (talker) |
| --- | --- | --- |
| X100000 | space | @ |
| X100001 | ! | A |
| X100010 | " | B |
| X100011 | #R | C |
| X100100 | $ | D |
| X100101 | % | E |
| X100110 | & | F |
| X100111 | ' | G |
| X101000 | ( | H |
| X101001 | ) | I |
| X101010 | ♯ | J |
| X101011 | + | K |
| X101100 | , | L |
| X101101 | − | M |
| X101110 | . | N |
| X101111 | / | O |
| X110000 | 0 | P |
| X110001 | 1 | Q |
| X110010 | 2 | R |
| X110011 | 3 | S |
| X110100 | 4 | T |
| X110101 | 5 | U |
| X110110 | 6 | V |
| X110111 | 7 | W |
| X111000 | 8 | X |
| X111001 | 9 | Y |
| X111010 | : | Z |
| X111011 | ; | [ |
| X111100 | < | \ |
| X111101 | = | ] |
| X111110 | > | ^ |

*Note:* bit 7 is 0 for listener addresses, 1 for talker addresses

to carry out a given command. Further information is supplied as another interface message, known as a **secondary command**. A secondary command facility effectively allows currently undefined commands to be included as and when required.

In a similar manner, **secondary addresses** are addresses which immediately follow a listen address or a talk address. Secondary addresses thus allow devices to be sub-divided into groups which may (or may not) be associated together as particular categories of device. So, for example, all printers may be grouped together in listen address group 1, all multimeters in talk address group 1, all signal generators in listen address group 2, all spectrum analysers in talk address group 2, and so on.

Relationships of commands and addresses, in terms of placings in the ASCII code are shown in Table 7.6, where interface messages (MSG) are shown directly adjacent (to the right of) their corresponding ASCII coded characters or control codes. Bits $b_1$ to $b_7$ are represented by logic states of the data bus DIO1 to DIO7. Command *parallel poll configure* (PPC) – see later in this chapter – requires following secondary commands from the secondary command group of Table 7.6.

### Other interface messages

Apart from controller commands there are many other interface messages used between controller and devices, or between devices.

Interface messages are listed in Table 7.7 as mnemonics, message name, type class and signal line logic levels. Type U refers to a uniline message, while type M refers to a multiline message. Classes of interface message are referred to as abbreviations, according to the following:

| Class | Meaning |
|---|---|
| AC | addressed command |
| AD | address (talk or listen) |
| DD | device-dependent |
| HS | handshake |
| SE | secondary |
| ST | status |
| UC | universal command |

Generally, letters in the data bus DIO1 to DIO7 signal lines refer to device-dependent bits, where letters D, E, L, T, S refer to data, *end of signal* code, listen address, talk address, and secondary address.

**Note** ⋆ specifies parallel poll response, where S refers to parallel poll response:

| S | Response |
|---|---|
| 0 | 0 |
| 1 | 1 |

**Table 7.6** Relationships of general-purpose interface bus commands and addresses, in terms of placings in the ASCII code

| $b_7$ $b_6$ $b_5$ | | 000 | 001 | 010 | 011 | 100 | 101 | 110 | 111 |
|---|---|---|---|---|---|---|---|---|---|
| $b_4$ $b_3$ $b_2$ $b_1$ | Row / Column | 0 | 1 | 2 | 3 | 4 | 5 | 6 | 7 |
| 0000 | 0 | NUL | DLE | SP | 0 | @ | P | ` | p |
| 0001 | 1 | SOH (GTL) | DC1 (LLO) | ! | 1 | A | Q | a | q |
| 0010 | 2 | STX | DC2 | " | 2 | B | R | b | r |
| 0011 | 3 | ETX | DC3 | # | 3 | C | S | c | s |
| 0100 | 4 | EOT (SDC) | DC4 (DCL) | $ | 4 | D | T | d | t |
| 0101 | 5 | ENQ (PPC) | NAK (PPU) | % | 5 | E | U | e | u |
| 0110 | 6 | ACK | SYN | & | 6 | F | V | f | v |
| 0111 | 7 | BEL | ETB | ' | 7 | G | W | g | w |
| 1000 | 8 | BS (GET) | CAN (SPE) | ( | 8 | H | X | h | x |
| 1001 | 9 | HT (TCT) | EM (SPD) | ) | 9 | I | Y | i | y |
| 1010 | 10 | LF | SUB | * | : | J | Z | j | z |
| 1011 | 11 | VT | ESC | + | ; | K | [ | k | { |
| 1100 | 12 | FF | FS | , | < | L | \ | l | \| |
| 1101 | 13 | CR | GS | - | = | M | ] | m | } |
| 1110 | 14 | SO | RS | . | > | N | ^ | n | ~ |
| 1111 | 15 | SI | US | / | ? (UNL) | O | _ (UNT) | o | DEL |

Column 0: Addressed command group (ACG)

Column 1: Universal command group (UCG)

Columns 2–3: Listen address group (LAG) — MLA ASSIGNED TO DEVICE ← → ; UNL

Columns 4–5: Talk address group (TAG) — MTA ASSIGNED TO DEVICE ← → ; UNT

Columns 0–5: Primary command group (PCG)

Columns 6–7: Secondary command group (SCG) — MEANING DEFINED BY PCG CODE ← →

**Table 7.7**  General-purpose interface bus interface messages, mnemonics, types, classes and codes

| Mnemonic | Message name | Type | Class | DIO 8 7 | 6 5 4 | DIO 3 2 1 | DAV NRFD NDAC | ATN | EOI | SRQ | IFC | REN |
|---|---|---|---|---|---|---|---|---|---|---|---|---|
| ACG | Addressed command group | M | AC | Y0 | 00X | XXX | XXX | 1 | X | X | X | X |
| ATN | Attention | U | UC | XX | XXX | XXX | XXX | 1 | X | X | X | X |
| DAB | Data byte | M | DD | DD (87) | DDD (654) | DDD (321) | XXX | 0 | X | X | X | X |
| DAC | Data accepted | U | HS | XX | XXX | XXX | XX0 | X | X | X | X | X |
| DAV | Data valid | U | HS | XX | XXX | XXX | 1XX | X | X | X | X | X |
| DCL | Device clear | M | UC | Y0 | 010 | 100 | XXX | 1 | X | X | X | X |
| END | End | U | ST | XX | XXX | XXX | XXX | 0 | 1 | X | X | X |
| EOS | End of string | M | DD | EE (87) | EEE (654) | EEE (321) | XXX | 0 | X | X | X | X |
| GET | Group execute trigger | M | AC | Y0 | 001 | 000 | XXX | 1 | X | X | X | X |
| GTL | Go to local | M | AC | Y0 | 000 | 001 | XXX | 1 | X | X | X | X |
| IDY | Identify | M | UC | XX | XXX | XXX | XXX | 1 | 1 | X | X | X |
| IFC | Interface clear | U | UC | XX | XXX | XXX | XXX | X | X | X | 1 | X |
| LAG | Listen address group | M | AD | Y0 | 010 | XXX | XXX | 1 | X | X | X | X |
| LLO | Local lockout | M | UC | Y0 | 010 | 001 | XXX | 1 | X | X | X | X |
| MLA | My listen address | M | AD | Y0 | 1LL (54) | LLL (321) | XXX | 1 | X | X | X | X |
| MTA | My talk address | M | AD | Y1 | 0TT (54) | TTT (321) | XXX | 1 | X | X | X | X |
| MSA | My secondary address | M | SE | Y1 | 1SS (54) | SSS (321) | XXX | 1 | X | X | X | X |
| NUL | Null byte | M | DD | 00 | 000 | 000 | XXX | X | X | X | X | X |
| OSA | Other secondary address | M | SE | (OSA = SCG∧MSA) | | | | | | | | |
| OTA | Other talk address | M | AD | (OTA = TAG∧MTA) | | | | | | | | |
| PCG | Primary command group | M | — | (PCG = ACG∨UCG∨LAG∨TAG) | | | | | | | | |

| Mnemonic | Message name | | | DIO 8 7 6 5 4 3 2 1 | DAV NRFD NDAC | ATN | EOI | SRQ | IFC | REN |
|---|---|---|---|---|---|---|---|---|---|---|
| PPC | Parallel poll configure | M | AC | Y0 000 101 | XXX | 1 | X | X | X | X |
| PPE | Parallel poll enable | M | SE | Y1 10S $P_3P_2P_1$ | XXX | 1 | X | X | X | X |
| PPD | Parallel poll disable | M | SE | Y1 11$D_4$ $D_3D_2D_1$ | XXX | 1 | X | X | X | X |
| PPR1 | Parallel poll response 1 | U | ST | XX XXX XX1 | XXX | 1 | 1 | X | X | X |
| PPR2 | Parallel poll response 2 | U | ST | XX XXX X1X | XXX | 1 | 1 | X | X | X |
| PPR3 | Parallel poll response 3 | U | ST | XX XXX 1XX | XXX | 1 | 1 | X | X | X |
| PPR4 | Parallel poll response 4 | U | ST | XX XX1 XXX | XXX | 1 | 1 | X | X | X |
| PPR5 | Parallel poll response 5 | U | ST | XX X1X XXX | XXX | 1 | 1 | X | X | X |
| PPR6 | Parallel poll response 6 | U | ST | XX 1XX XXX | XXX | 1 | 1 | X | X | X |
| PPR7 | Parallel poll response 7 | U | ST | X1 XXX XXX | XXX | 1 | 1 | X | X | X |
| PPR8 | Parallel poll response 8 | U | ST | 1X XXX XXX | XXX | 1 | 1 | X | X | X |
| PPU | Parallel poll unconfigure | M | UC | Y0 010 101 | XXX | 1 | X | X | X | X |
| REN | Remote enable | U | UC | XX XXX XXX | XXX | X | X | X | X | 1 |
| RFD | Ready for data | U | HS | XX XXX XXX | X0X | X | X | X | X | X |
| RQS | Request service | U | ST | X1 XXX XXX | XXX | 0 | X | X | X | X |
| SCG | Secondary command group | M | SE | Y1 1XX XXX | XXX | 1 | X | X | X | X |
| SDC | Selected device clear | M | AC | Y0 000 100 | XXX | 1 | X | X | X | X |
| SPD | Serial poll disable | M | UC | Y0 011 001 | XXX | 1 | X | X | X | X |
| SPE | Serial poll enable | M | UC | Y0 011 000 | XXX | 1 | X | X | X | X |
| SRQ | Service request | U | ST | XX XXX XXX | XXX | X | X | 1 | X | X |
| STB | Status byte | M | ST | $S_8$X $S_6S_5S_4$ $S_3S_2S_1$ | XXX | 0 | X | X | X | X |
| TCT | Take control | M | AC | Y0 001 001 | XXX | 1 | X | X | X | X |
| TAG | Talk address group | M | AD | Y1 0XX XXX | XXX | 1 | X | X | X | X |
| UCG | Universal command group | M | UC | Y0 01X XXX | XXX | 1 | X | X | X | X |
| UNL | Unlisten | M | AD | Y0 111 111 | XXX | 1 | X | X | X | X |

and P1 to P3 refers to *parallel poll response* (PPR) message to be sent when a parallel poll is executed:

| P3 | P2 | P1 | PPR *message* |
|----|----|----|---------------|
| 0  | 0  | 0  | PPR1          |
| 0  | 0  | 1  | PPR2          |
| .  | .  | .  | .             |
| .  | .  | .  | .             |
| 1  | 1  | 1  | PPR8          |

**Note** ** refers to status byte format, where S1 to S6 and S8 specify device-dependent status – see later in this chapter.

## Command sequences

Most communications facilities over the general-purpose interface bus require sequences of interface messages, which are simply combinations of multiline and uniline interface messages. In a large number of instances these sequences follow similar patterns, which may be standardized to ensure conformity in controlling software. There are, consequently, a few operational sequences of commands which are used so regularly they may be considered almost controller commands in their own right. Some of the more common are:

- Data handshake.
- Data transfer.
- Serial poll.
- Parallel poll.
- Control passing.
- Enforcing remote control.

Command sequences which allow these communications facilities follow. It is important to note these sequences are not the only ways of providing the facilities, however.

### Data handshake

The most important of these command sequences is a foolproof system whereby data can be transferred correctly between devices, despite differing data speeds which devices connected to the general-purpose interface bus may have.

Transferring data between one device and another using this **data handshake** sequence is a reasonably simple task. The controller first assigns one device as a talker and one or more devices as listeners, issuing talker and listener addresses over the bus. To do this the controller *attention* (ATN)

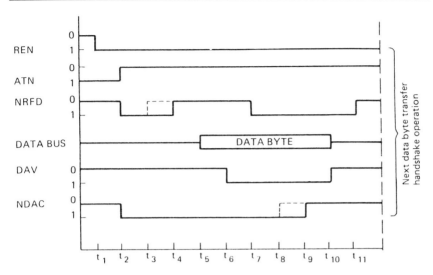

**Figure 7.4** *Timing diagram of signals illustrating the handshake procedure between a talker and listeners, required to allow asynchronous transmission of data across the general-purpose interface bus*

signal line is first set to logic 1, thus informing all devices the data bus contains address information. After talker and listener addresses have been issued the controller changes the *attention* signal line to logic 0, and data transfer control is passed to the talker. The talker now simply places data onto the bus and listeners simply accept data in an asynchronous handshaking procedure, one byte at a time.

Figure 7.4 shows a timing diagram which illustrates the basic handshake procedure between talker and listeners on the general-purpose interface bus (after the controller has assigned one device as talker, and one or more devices as listeners).

The controller first sends a *remote enable* (REN at $t_1$) and signals *attention* (ATN at $t_2$), which causes devices to set their *not ready for data* (NRFD) and *not data accepted* (NDAC) lines. At $t_3$ the fastest device is ready to accept data. However, as the *not ready for data* (NRFD) signals of all connected devices are in AND logic, line logic level only changes when *all* listener devices are ready to accept data (NRFD at $t_4$).

The talker now presents the data byte on data bus lines DIO1 to DIO8 (at $t_5$); then indicates this with a *data valid* (DAV at $t_6$) signal. Devices reset the *not ready for data* (NRFD at $t_7$) signal.

At $t_8$ the fastest device has accepted the data byte, but the *not data accepted* (NDAC) signal line is only reset when *all* listeners have accepted the data byte, at $t_9$.

The talker now removes the data byte and clears the *data valid* (DAV) signal, at $t_{10}$. After an interval, which depends on the time each listening devices needs to process the data byte, the *not ready for data* (NRFD at $t_{11}$) signal changes once again and the data transfer proceeds.

As three signal lines are used for this handshake; *data valid* (DAV), *not ready for data* (NRFD) and *not data accepted* (NDAC), the procedure is often called a **three-line handshake**. It provides a simple but effective method of transferring data bytes over the general-purpose interface bus, with the advantage that it does not take up much of the controller's processing time.

### Data transfer

Although a data handshake ensures correct data transfer between devices with different data speeds, where devices have the same data speeds it is often not essential. A much simpler data transfer command sequence is then usually preferred. Initially, the controller issues an *attention* (ATN) signal by taking the *attention* line to logic 1, then issues an *unlisten* (UNL) message to turn off all listening devices. Listen address (LAD)$_1$ to (LAD)$_n$ are issued, which specify which devices are to receive following data bytes. The controller then issues *talk address* (TAD), then takes the *attention* line to logic 0.

The specified talker proceeds to transmit data bytes (DAB)$_1$ to (DAB)$_n$, which are received by all specified listeners. Data transfer ends when the talker takes the *end of identify* (EOI) signal line to logic 1, or sends an *end of signal* (EOS) message, or the controller takes the *attention* line to logic 1 again. This may be summarized:

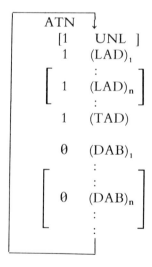

## Serial poll

Where individual devices need to attract the controller's attention, polling sequences are used. There are two types of polling sequence: first is serial poll, which uses the *service request* (SRQ) interface management line of the general-purpose interface bus to attract the controller's attention. A logic 1 signal initiates the poll. First, the controller issues an *attention* (ATN) signal, together with an *unlisten* (UNL) message which prevents all devices from listening to serial poll data. Next, the controller sends a *serial poll enable* (SPE) message over the data bus.

Next, the controller issues *talk address* (TAD) of the first device on the bus, then takes the *attention* line to logic 0. The specified device returns a **status byte** (STB) in the form (SBN) or (SBA), which indicates its operating condition to the controller. Status byte format is given in Table 7.8.

Once this status byte is received, the controller takes *attention* line to logic 1 again, issues *talk address* of the next device and accepts its status byte. This **polling** procedure is followed until all devices on the general-purpose interface bus have been polled, after which the controller issues a *serial poll disable* (SPD) universal command, followed by an *untalk* (UNT) command (to disable the last polled device from talking as soon as the attention line goes to logic 0).

Aspects of the serial poll procedure are adjustable by software. For example, the order of devices polled may be defined by operator commands in the controlling program. Similarly, controller response to a poll may be dependent on which device polls it.

The most important feature of a general-purpose interface bus serial poll procedure is the fact that devices initiate the serial poll sequence by

**Table 7.8** Status byte format, returned after a serial poll of the general-purpose interface bus

| Condition | | | Data lines | | | | | |
|---|---|---|---|---|---|---|---|---|
| | 8 | 7 (RQS) | 6 | 5 | 4 | 3 | 2 | 1 |
| 1 | X or E | service requested | abnormal alarm | busy | X | X | X | X |
| 0 | X or E | service not requested | conditions normal | ready | X | X | X | X |

*Notes:* X indicates data which may be specific to an instrument
E indicates use of a possible extension to status message, in data lines 1 to 4

attracting the controller's attention with the *service request* (SRQ) line, only taking up controller time when attention is needed.

Serial poll sequence may be summarized:

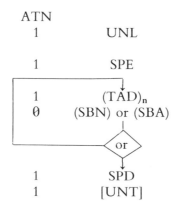

ATN
| 1 | UNL |
|---|---|
| 1 | SPE |
| 1 | $(TAD)_n$ |
| 0 | (SBN) or (SBA) |
| 1 | SPD |
| 1 | [UNT] |

### Parallel poll

The second type of polling procedure is **parallel poll**, which is generally executed by the controller of a general-purpose interface bus periodically – unlike a serial poll which is initiated by a device's request for service. Parallel polled devices have no facility for attracting the controller's attention.

Before a controller may use the parallel polling procedure it must allocate *service request* lines on the data bus. Allocation is announced by the controller by first calling all devices' attention by taking the *attention* line to logic 1, stating the *listen address* of a particular device (LAD) and issuing a *parallel poll configure* (PPC) command followed with a *parallel poll enable* (PPE) codeword, which incorporates a 3-bit code (bits 1 to 3) signifying which of the eight data lines is to be used to signify device status. If the device requires attention it must hold its designated data line at the logic level specified by bit 4 of the *parallel poll enable* (PPE) message issued. Codeword and sense are memorized by the device.

This may be summarized:

| ATN | IDY | |
|---|---|---|
| 1 | 0 | $(LAD)_n$ |
| 1 | 0 | PPC |
| 1 | 0 | PPE |
| 1 | 0 | UNL |

Allocation continues for each device required to be included in a parallel poll sequence.

The parallel poll sequence itself takes place as the controller takes the bus *attention* (ATN) line to logic 1 and issues an *identify* (IDY) command. Data lines held at specified logic levels by devices during this period indicate which devices require service. This may be summarized:

ATN   IDY
1        1

This procedure allows up to eight devices to be uniquely identified by the controller in a single rapid polling operation. Further, it is acceptable to allocate each data line to more than one device for parallel polling purposes, so all devices on a general-purpose interface bus system can be polled in a parallel poll – although the controller must then identify which of the two or more devices signalling on a specific data line is requesting controller attention.

Although a parallel polling sequence does not allow devices to initiate a request for service, it is significantly faster than a serial polling sequence. Periodic parallel polling, on the other hand, must take place (and be written into the controlling program) otherwise devices are not serviced.

## Control passing

Occasionally a specified controller may pass control to another device, which then becomes system controller. After taking *attention* line (ATN) to logic 1 to gain all devices' attention, the controller issues the *talk address* (TAD) of the new controller, then issues the *take control* (TCT) command, and takes *attention* line to logic 0, at which time the new controller is in charge. This may be summarized:

ATN
1              (TAD)

1              TCT
0              –

## Enforcing remote control

One of the main features of the general-purpose interface bus is that devices' front panel controls are controllable by the system controller. This is enforced by a controller after taking *attention* (ATN) and *remote enable* (REN) lines to logic 1, by issuing a *local lockout* (LLO) message. Then, by issuing *listener addresses* $(LAD)_1$ to $(LAD)_n$, addressed devices are forced into remote control. This may be summarized:

| ATN | REN | |
|---|---|---|
| 1 | 1 | LLO |
| 1 | 1 | $(LAD)_n$ |

## Interface messages and functions

Interface functions call up interface messages in set ways. Table 7.9 lists interface messages as mnemonics, together with message names and those interface functions which call them. For the sake of completeness, local messages are listed similarly in Table 7.10.

**Table 7.9** General-purpose interface bus interface messages as mnemonics and those interface functions which call them

| Mnemonic | Message | Interface function |
|---|---|---|
| *Received* | | |
| ATN | Attention | SH, AH, T, TE, L, LE, PP, C |
| DAB | Data byte | (via L, LE) |
| DAC | Data accepted | SH |
| DAV | Data valid | AH |
| DCL | Device clear | DC |
| END | End | (via L, LE) |
| GET | Group execute trigger | DT |
| GTL | Go to local | RL |
| IDY | Identify | PP, L, LE |
| IFC | Interface clear | T, TE, L, LE, C |
| LLO | Local lockout | RL |
| MLA | My listen address | L, LE, RL |
| [MLA] | My listen address | T |
| MSA or [MSA] | My secondary address | TE, LE |
| MTA | My talk address | T, TE |
| [MTA] | My talk address | L |
| OSA | Other secondary address | TE |
| OTA | Other talk address | T, TE |
| PCG | Primary command group | TE, LE, PP |
| PPC | Parallel poll configure | PP |
| [PPD] | Parallel poll disable | PP |
| [PPE] | Parallel poll enable | PP |
| PPRn | Parallel poll response n | (via C) |
| PPU | Parallel poll unconfigure | PP |
| REN | Remote enable | RL |
| RFD | Ready for data | SH |

**Table 7.9** *(cont.)*

| Mnemonic | Message | Interface function |
|---|---|---|
| RQS | Request service | (via L, LE) |
| [SDC] | Selected device clear | DC |
| SPD | Serial poll disable | T, TE |
| SPE | Serial poll enable | T, TE |
| SRQ | Service request | (via C) |
| STB | Status byte | (via L, LE) |
| TCT or [TCT] | Take control | C |
| UNL | Unlisten | L, LE |
| *Sent* | | |
| ATN | Attention | C |
| DAB | Data byte | (via T, TE) |
| DAC | Data accepted | AH |
| DAV | Data valid | SH |
| DCL | Device clear | (via C) |
| END | End | (via T) |
| GET | Group execute trigger | (via C) |
| GTL | Go to local | (via C) |
| IDY | Identify | C |
| IFC | Interface clear | C |
| LLO | Local lockout | (via C) |
| MLA or [MLA] | My listen address | (via C) |
| MSA or [MSA] | My secondary address | (via C) |
| MTA or [MTA] | My talk address | (via C) |
| OSA | Other secondary address | (via C) |
| OTA | Other talk address | (via C) |
| PCG | Primary command group | (via C) |
| PPC | Parallel poll configure | (via C) |
| [PPD] | Parallel poll disable | (via C) |
| [PPE] | Parallel poll enable | (via C) |
| PPRn | Parallel poll response n | PP |
| PPU | Parallel poll unconfigure | (via C) |
| REN | Remote enable | C |
| RFD | Ready for data | AH |
| RQS | Request service | T, TE |
| [SDC] | Selected device clear | (via C) |
| SPD | Serial poll disable | (via C) |
| SPE | Serial poll enable | (via C) |
| SRQ | Service request | SR |
| STB | Status byte | (via T, TE) |
| TCT | Take control | (via C) |
| UNL | Unlisten | (via C) |

**Table 7.10**   General-purpose interface bus local messages as mnemonics and those interface functions which call them

| Mnemonic | Message | Interface functions |
|---|---|---|
| *Received by interface functions* | | |
| gts | go to standby | C |
| ist | individual service request (qualifier) | PP |
| lon | listen only | L, LE |
| [lpe] | local poll enable | PP |
| ltn | listen | L, LE |
| lun | local unlisten | L, LE |
| nba | new byte available | SH |
| pon | power on | SH, AH, T, TE, L, LE SR, RL, PP, C |
| rdy | ready | AH |
| rpp | request parallel poll | C |
| rsc | request system control | C |
| rsv | request service | SR |
| rtl | return to local | RL |
| sic | send interface clear | C |
| sre | send remote enable | C |
| tca | take control asynchronously | C |
| tcs | take control synchronously | AH, C |
| ton | talk only | T, TE |
| *Sent by interface functions* | | |

## Device-dependent messages

Device-dependent messages are, of course, irrelevant to general-purpose interface bus operation. They are transmitted between devices after a transmission path has been set up by interface messages and do not affect the bus itself directly. Consequently their formats are irrelevant, too.

Generally, however, in common with all networked information transfer systems, messages transmitted over the general-purpose interface bus typically have standard formats, suggested in worldwide standards as guidelines and alternatives. These formats are more recommendations than standards, though conformance does greatly ease programming requirements and use, and tends to reduce idiosyncrasies exhibited by particular devices.

There are four main formats of device-dependent messages, differing mainly because they are used for four corresponding messages types. Main device-dependent message types are:

- measurement data
- display data
- program or control data
- status data.

Message formats of these four main message types may vary from device to device so it is essential they are adequately defined by manufacturers, so programming may take differences into account.

### General format

Typical device-dependent message format comprises a number of fields made up of data bytes: a **header** of alpha data; a **body** of numeric data; an **ending**, known as a **delimiter**, shown in Figure 7.5.

| Header | Body | Delimiter |
|--------|------|-----------|

**Figure 7.5** *Typical format of a device-dependent message of the general-purpose interface bus*

Messages may be individual, or may be grouped together to form a sequential chain of message units (see later). Note that some device-dependent messages – device status (STB) and end (END) – are single-byte or single-bit messages: these have previously been described here as interface messages, defined as part of the general-purpose interface bus specification. Further, not all device-dependent messages need (or use) all fields of this general format.

A common method of identifying individual data fields uses letters, according to:

| Letter | Field contents | Field |
|--------|----------------|-------|
| T | Type and/or quality of data | Header (alpha) |
| U | Sign or polarity of data | |
| V | Numeric value of data | Body (numeric) |
| W | Exponent notation | |
| X | String delimiter | |
| Y | Block delimiter | Ending (delimiter) |
| Z | Record delimiter | |

Fields of message unit are transmitted by a device in the sequence from left to right:

$$\boxed{T} \ \boxed{U} \ \boxed{V} \ \boxed{W} \ \boxed{X} \ \boxed{Y} \ \boxed{Z}$$

Within each field of the message, the most significant digit or character is transmitted first.

*Data field* [T] is used to describe type and quality of the numeric data. It should be limited to alpha data and should be as short as possible, and preferably of fixed length for each unique product. If the field refers to units, unscaled units (such as V, A, F, Ω) are preferred, though scaling in accordance with IEC Publication 27 (kV, mA, μF, MΩ) is allowed.

*Sign field* [U] is limited to a single character to indicate the sign of following numeric data.

*Numeric value field* [V] is of variable length to fit particular applications.

*Exponent field* [W] is used to give the exponent of the numerical value and comprises three parts in the sequence:

- Character E, indicating subsequent numbers, including sign, form an exponent to the base 10.
- + or −, indicating exponent sign. If the exponent is zero, + is used.
- Characters (two-digit preferred) indicating exponent value.

*Delimiter fields* [X], [Y], and [Z] are used to delimit message units of single or multiple form (see later in this chapter).

### Measurement data and display data

To most intents and purposes messages of measurement data and display data have the same format as the preceding sequence of fields. All fields apart from the numeric value held in the [V] field are entirely optional, depending on application alone. Data field [T], numeric value field [V] and exponent notation field [W] are of variable length.

### Program data

Program data messages typically comprise a series of message units, each of which is constructed from [T] and [V] fields, together with required delimiters. Numeric field [V] is, however, optional. Further, suitable choices of [T] and [V] fields can reduce (or even eliminate) the need for delimiters between message units.

### Status data

Messages of status data are simply status bytes, listed earlier in Table 7.8. However, status messages may contain considerable other device-dependent information, which is defined by the manufacturer depending on application, device, and requirement.

**Table 7.11** Device-dependent message fields T, U, V and W, together with their interpretations, use, number of bytes, characters and codes

| Field | Interpretation | Measurement data | Program data | Number of bytes when used | Characters Preferred | Characters Allowed | Comments |
|---|---|---|---|---|---|---|---|
| T | Unit or quality | O | M | As small as possible | Any alpha characters from columns 4 and 5 | Any alpha characters from columns 6 and 7, and Δ (space) | Use E with caution (it is used in field W) Use non-alpha characters from columns 2 to 5 with caution for program data (+|−|.|; and digits 0 to 9 are not allowed) |
| U | Sign | O | O | 1 | \|+\|−\| | Δ (space) | No spaces at beginning of V field if U field is + or − |
| V | Numerals | M | O | ≥1 | Any numerals from column 3 | Δ (space) | No embedded or trailing spaces |
| W | Decimal point | O | O | 1 | \|.\|(period) | No others | |
| | Identifier | | | 1 | E | No others | The W field is either to be used completely or not to be used at all |
| | Exponent notation Sign | | M | 1 | \|+\|−\| | No others | |
| | Numerals | | | ≤2 | Any numerals from column 3 | No others | Preferred: two digits Allowed: one digit |

**Table 7.12**   ISO 7-bit code, used for characters in device-dependent message fields of the general-purpose interface bus

| | $b_7$ | 0 | 0 | 0 | 0 | 1 | 1 | 1 | 1 |
|---|---|---|---|---|---|---|---|---|---|
| | $b_6$ | 0 | 0 | 1 | 1 | 0 | 0 | 1 | 1 |
| | $b_5$ | 0 | 1 | 0 | 1 | 0 | 1 | 0 | 1 |
| | Column | | | | | | | | |
| $b_4\ b_3\ b_2\ b_1$ Row | | 0 | 1 | 2 | 3 | 4 | 5 | 6 | 7 |
| 0 0 0 0 | 0 | NUL | TC$_7$ (DLE) | SP | 0 | @ | P | | p |
| 0 0 0 1 | 1 | TC$_1$ (SOH) | DC$_1$ | ! | 1 | A | Q | a | q |
| 0 0 1 0 | 2 | TC$_2$ (STX) | DC$_2$ | " | 2 | B | R | b | r |
| 0 0 1 1 | 3 | TC$_3$ (ETX) | DC$_3$ | ♯ | 3 | C | S | c | s |
| 0 1 0 0 | 4 | TC$_4$ (EOT) | DC$_4$ | ¤ | 4 | D | T | d | t |
| 0 1 0 1 | 5 | TC$_5$ (ENQ) | TC$_8$ (NAK) | % | 5 | E | U | e | u |
| 0 1 1 0 | 6 | TC$_6$ (ACK) | TC$_9$ (SYN) | & | 6 | F | V | f | v |
| 0 1 1 1 | 7 | BEL | TC$_{10}$ (ETB) | ' | 7 | G | W | g | w |
| 1 0 0 0 | 8 | FE$_0$ (BS) | CAN | ( | 8 | H | X | h | x |
| 1 0 0 1 | 9 | FE$_1$ (HT) | EM | ) | 9 | I | Y | i | y |
| 1 0 1 0 | 10 | FE$_2$ (LF) | SUB | * | : | J | Z | j | z |
| 1 0 1 1 | 11 | FE$_3$ (VT) | ESC | + | ; | K | [ | k | { |
| 1 1 0 0 | 12 | FE$_4$ (FF) | IS$_4$ (FS) | , | < | L | \ | l | | |
| 1 1 0 1 | 13 | FE$_5$ (CR) | IS$_3$ (GS) | – | = | M | ] | m | } |
| 1 1 1 0 | 14 | SO | IS$_2$ (RS) | . | > | N | ^ | n | – |
| 1 1 1 1 | 15 | SI | IS$_1$ (US) | / | ? | O | – | o | DEL |

☐ Coding for NL is the same as for LF (FE$_2$).

## Device-dependent message codes

In summary, codes for use in device-dependent message units in fields [T], [U], [V], and [W] are given in Table 7.11. In measurement data and program data columns of the table, O and M refer to optional and mandatory use. A vertical bar (|) is used to separate specified characters. Characters in individual fields are from the ISO 7-bit code, which is given in Table 7.12 for reference.

## Delimiters

Message units may be joined sequentially in structures known as files, records, blocks and strings to create longer messages. Further, devices used in instrumentation often transmit data in a continuous form. To allow receiving devices to distinguish between message parts in a message structure, delimiters are used. There are three types of delimiters, identified as fields within a message unit: where the [X] field indicates end of a message string; the [Y] field indicates end of a message block; the [Z] field indicates end of a message record, according to the message structure format shown in Figure 7.6. Delimiters must not be consecutive, apart from the only exception of block and record delimiters.

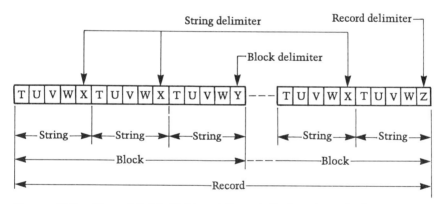

**Figure 7.6** *Use of fields X, Y and Z as delimiters in a device-dependent message structure of the general-purpose interface bus*

Table 7.13 lists delimiters, types, codes and applications. If a record consists of a single block, no additional block delimiter is required. If a block consists of a single string, no additional string delimiter is required.

## Devices

While implementation of a general-purpose interface bus interface within a particular device requires device functions, interface functions, message

**Table 7.13**  Delimiters allowed in fields X, Y and Z of general-purpose interface bus device-dependent marriages; listing fields, delimiter types, codes and comments

| Field | Delimiter type | ISO 7-bit code | Comments |
|---|---|---|---|
| X | String | , (preferred)<br>; (for a higher level) | String delimiters used for:<br>• sequential message units of the *same type*, repeated measurements such as $UV_1$, $UV_1$, $UV_1$ ... |
| Y | Block | *NL* (preferred)<br>*ETB*<br>*CR LF* | Block delimiters used for:<br>• single measured value message unit such as TUV *NL*<br>• related message unit ending such as $TUV_1$, $UV_2$, $UV_3$, *NL* |
| Z | Record | END (an interface message on the EOI line) | Record delimiters used for:<br>• conclusion of one or more blocks of data where the talker function requires further instructions before output of another message sequence such as $TUV_1$, $TUV_2$ ... TUDDD.DD $[D = DAB]$, END sent concurrent with last DAB |

coding logic, together with requisite bus drivers and receivers similar to those shown in Figure 7.3, it must be remembered these are all logical parts; which may not have individual hardware parts in a finished device's interface.

Interfaces are responsible for three main tasks within the general-purpose interface bus system:

• Interface functions.
• Interface message communication.
• Interface message coding and decoding.

These three tasks are well defined and, consequently, interfaces to the general-purpose interface bus are incorporated into devices in a fairly set manner, which may be generalized quite easily. Although specific devices differ, layout of a typical device is given as a block diagram in Figure 7.7. Here device control is usually of a microprocessor-based form which often, in fact, assumes control of all parts of the device. Usually, the interface to the general-purpose interface bus is controlled by a single separate microprocessor, effectively under microprocessor-based device control though operating autonomously under most circumstances. Interface drivers and receivers may be located on separate integrated circuits.

## Size and speed limitations

The general-purpose interface bus has two limitations which should be borne in mind when constructing an automatic test equipment system. First, the maximum number of instruments, including the controller, which can be connected to a basic bus is 15 (although with extended addressing – see earlier – this may be extended). The maximum cable length is calculated as:

2 metres times the number of instruments

or:

20 metres

whichever is less. The maximum distance between any two instruments is four metres.

With a 20 metre cable, the bus can operate with a data transfer rate of up to 250,000 byte $s^{-1}$, but with a shorter cable length considerably higher data transfer rates (up to 1,000,000 byte $s^{-1}$) can be achieved. In most practical applications, however, overall data transfer rate is much less than these maxima, being limited by the speed of operation of the instruments connected, rather than the bus itself.

An extender may be used to allow use of a general-purpose interface bus system over greater distances than the 20 m limit. Typically, these may incorporate CCITT V24 (IEEE EIA232) serial links over telephone lines, or optical fibre links.

## Hardware

As far as a user is concerned, general-purpose interface bus hardware comprises nothing more than a number of lengths of cable terminated at each end with a connector. Connecting up a general-purpose interface bus system is a simple matter of plugging in the connectors to mating connectors at the rear of all the required test instruments and other devices.

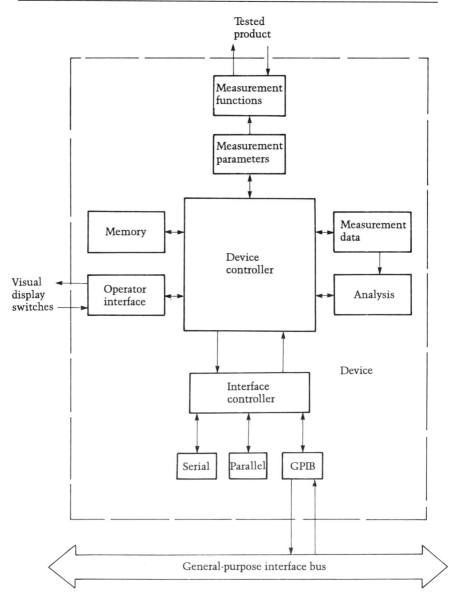

**Figure 7.7** *Block diagram of general-purpose interface bus device*

This is usually by daisy-chaining leads, from one instrument to the next, until all system components are connected. Connectors are usually stackable, so one connector plugs into the back of another, as shown in Figure 7.8. Where test instruments to the different international standards IEC625 and IEEE488 are used commonly available adaptors are required, as connectors are slightly different.

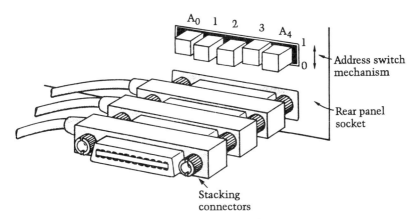

**Figure 7.8** *Stacked general-purpose interface bus connectors*

IEC625 connectors for the general-purpose interface bus have 25 connection pins, while IEEE488 connectors have 24 pins, the reason being that IEC625 allows for an additional ground return line. Pin assignments for the general-purpose interface bus are listed in Table 7.14, while Figure 7.9 illustrates pin numbering of the IEC625 connector and Figure 7.10 illustrates the IEEE488 connector.

To all practical intents, though, general-purpose interface buses of the two standards are otherwise compatible, and only a simple interface adaptor is required.

### Logical and electrical state relationships

The general-purpose interface bus operates in negative logic. In other words a logical coding state of 0 is represented by an electrical voltage of 5 volts (nominally $\geq +2$ V, called *high*), while a logical coding state of 1 is represented by a voltage of 0 volts (nominally $\leq 0.8$ V, called *low*). High and low states are based on standard TTL levels for which system power supplies do not exceed $+5.25$ V DC, and are referenced to logic ground.

To avoid confusion here, when referring to logical states of signal lines throughout this chapter, states are referred to only as logic 0 or log 1. The terms *high* and *low* are not used.

**Table 7.14**   Pin assignments of the general-purpose interface bus cable connector

| Pin no. | Signal group | Abbreviation | Signal/function |
|---|---|---|---|
| 1 | Data | D101 | Data line 1 |
| 2 |  | D102 | Data line 2 |
| 3 |  | D103 | Data line 3 |
| 4 |  | D104 | Data line 4 |
| 5 | Management | EOI | End or identify (sent by a talker to indicate that transfer of data is complete) |
| 6 | Handshake | DAV | Data valid (asserted by a talker to indicate that valid data is present on the bus) |
| 7 |  | NRFD | Not ready for data (asserted by a listener to indicate that it is not ready for data) |
| 8 |  | NDAC | Not data accepted (asserted while data is being accepted by a listener) |
| 9 | Management | IFC | Interface clear (asserted by the controller in order to initialize the system in known state) |
| 10 |  | SRQ | Service request (sent to the controller by a device requiring attention) |
| 11 |  | ATN | Attention (asserted by the controller when placing a command onto the bus) |
| 12 |  | SHIELD | Shield |
| 13 | Data | D105 | Data line 5 |
| 14 |  | D106 | Data line 6 |
| 15 |  | D107 | Data line 7 |
| 16 |  | D108 | Data line 8 |
| 17 | Management | REN | Remote enable (enables an instrument to be controlled by the bus controller rather than by its own front panel controls) |
| 18 |  | GND | Ground/common |
| 19 |  | GND | Ground/common |
| 20 |  | GND | Ground/common |
| 21 |  | GND | Ground/common |
| 22 |  | GND | Ground/common |
| 23 |  | GND | Ground/common |
| 24 |  | GND | Ground/common |

*Notes:* (a) Handshake signals (DAV, NRFD and NDAC) are all active low open collector and are used in a wired-OR configuration.

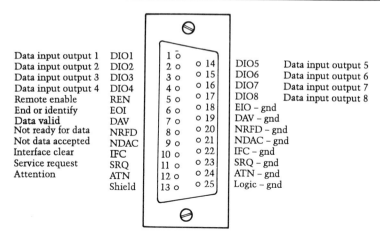

**Figure 7.9** *Pin numbering of IEC625 connector, viewed into female socket*

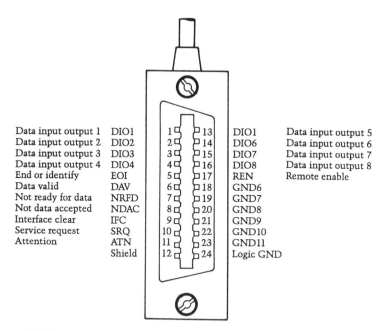

**Figure 7.10** *Pin numbering of IEEE488 connector, viewed into female socket*

## Drivers

Open collector drivers are used on *service request* (SRQ), *not ready for data* (NRFD) and *not data accepted* (NDAC) signal lines of the bus. Open collector *or* three-state drivers may be used on all other signal lines. However, where parallel polling is used, data bus DIO1 to DIO8 must comprise open collector drivers. Further, where devices have three-state drivers on data bus, *data valid* (DAV) and *end or identify* (EOI) signal lines it is recommended that the *attention* (ATN) line of the controller is also a three-state driver.

Driving a logic 1: output voltage of drivers must be no more than $+0.5$ V at $+48$ mA sink current, and drivers must be capable of sinking 48 mA continuously.

Driving a logic 0: output voltage of drivers must be at least $+2.4$ V at $-5.2$ mA.

## Receivers

Input voltages of receivers are no more than $+0.8$ V for a logic 1, and at least $+2$ V for logic 0. Schmitt-type receivers are recommended which give greater noise immunity.

Each signal line of a device is internally terminated by a resistive load. The main purpose of this resistive load is to establish a steady-state voltage even if all drivers on a line are in a high impedance state. A secondary purpose of this resistive termination is to give uniform device impedance on each line. Signal lines are clamped (usually with a diode clamping arrangement) which limits negative voltage excursions.

Internal capacitance load on each signal line is limited to a maximum of 100 pF within each device, substantially to ground.

## Interface cables

As all other interface functions are incorporated inside system components connected to the bus, interface cables themselves are passive. For this reason cables are often ignored when considering system performance. Far from this being the case, cables are quite important. Table 7.15 lists important electrical parameters of general-purpose interface bus cables; these are all worst case specifications.

Each signal line: *data valid* (DAV), *not ready for data* (NRFD), *not data accepted* (NDAC), *interface clear* (IFC), *attention* (ATN), *remote enable* (REN), *service request* (SRQ), and *end of identify* (EOI) forms a twisted-pair of wires with a logic ground wire, or isolated using an equivalent scheme, to minimize crosstalk. Sometimes cables which have twisted pairs for the remaining data bus lines DIO1 to DIO8 are used, too.

**Table 7.15** Electrical parameters of general-purpose interface bus cable

| Parameter | Value |
|---|---|
| Maximum resistance per metre length of inner conductors: | |
|     signal lines | $0.14\,\Omega$ |
|     signal ground returns | $0.14\,\Omega$ |
|     common logic ground return | $0.085\,\Omega$ |
|     outer screen | $0.0085\,\Omega$ |
| Maximum capacitance (at 1 kHz) between: | |
|     signal lines and all ground return lines | $150\,pF$ |
| Minimum density covering of braided shield | $85\,\%$ |

*Note:* cable should have an outer braided shield screen and at least 24 inner conductors, 16 of which are signal lines and the others are logic ground lines.

The overall shield of the interface cable is connected through one contact of the connector to frame potential, to minimize susceptibility to and generation of external noise. However, this does mean devices of significantly different frame potentials should not be used, as the interface cable may not be capable of handling excessive ground currents.

Ground returns of individual signal lines should be connected to logic ground at the logic circuit driver or receiver to minimize crosstalk interference transients. There should be a removable link connecting logic ground to frame potential if frame potential is connected to a metal device enclosure or the protective earth potential allowing a mechanism of avoiding ground loops.

# 8 VMEbus

One of the most important buses designed specifically for modular computer systems is the VMEbus. Like many of the bus systems already discussed in Chapter 6, it is a backplaned system, such that modular boards may be plugged into chassis-mounted connectors. Pin connections to connectors are such that a board may be plugged into the bus at any location within the chassis (slot 1, however, is reserved for the system controller board).

Although electrically and logically similar to the 68000 microprocessor-based computer system, VMEbus is an open system, not dependent on any particular microprocessor. VMEbus systems based on other micropro-cessors are, indeed, common.

Not by coincidence, however, VMEbus is a direct development of the 68000 microprocessor-based computer system. It is, in fact, a derivative of Motorola's **Versabus** computer bus system, originally defined in 1979 for the 68000 microprocessor. Motorola realized the potential of a bus which did not rely on a specific microprocessor, and so teamed with Mostek and Signetics to define VMEbus. Even the name VMEbus is indicative of its origins, and is simply a contraction of the term *Versabus on a modular Eurocard*.

Basic team strategies Motorola, Mostek and Signetics agreed were to:

- Design an all-purpose 16-bit/32-bit computer bus, to the best possible standard.
- Place the bus into public domain, without copyright – without licensing, without royalties.
- Establish an independent organization to provide stewardship and strong promotion of the bus.

VME bus was defined by the team in 1981, and immediately made available to all. Since then, VMEbus has been standardized in IEC821 and IEEE1014, and some substantial improvements have been incorporated in various revisions to these standards.

Stewardship of VMEbus is allocated to the *VMEbus International Trade Association* (VITA), which was formed in 1984 for the purpose. VITA is a non–profitmaking organization, with aims of ensuring technical and commercial success of VMEbus throughout the world. In these respects VITA is succeeding – VMEbus is reputed to be the industrial computer bus system *de facto* standard.

Although based on Versabus, and relying heavily on many of its fundamentals regarding signals, protocols and electrical specifications, VMEbus is independent of any particular microprocessor, being the first true **open architectured** bus suitable for 16-bit or 32-bit computer use (64-bit, in the case of the **VMEbus Futurebus + extended architecture** proposed in Revision D of the IEEE1014 standard – see later) we have considered. This open architectured, modular format of VMEbus gives a significant improvement in performance over other buses, and although not yet ideal for automatic test equipment systems, forms the basis of many similar computing opportunities. Nevertheless, automatic test equipment systems based on VMEbus *do* exist, usually on a turnkey basis (interestingly not – as readers may expect – modular) and are extremely good examples of VMEbus use. Where modular automatic test equipment systems are concerned, however, VXIbus forms a preferred specification; but VXIbus (described fully in Chapter 9) is, of course, simply a derivative of VMEbus, anyway.

## Hardware essentials

VMEbus board size is standardized into two main sizes based on single- and double-height Eurocards specified in standard IEC297.3. A single-height module (often referred to as a 3U module) has a single connector, known as the P1 connector. Double-height modules (often called 6U modules), on the other hand, must have the P1 connector with an optional P2 connector. Slot 1 in the chassis is assigned to a system controller, while the remainder of the 21 possible slots of the bus system may be taken by any combination of single- or double-height modules of whatever function.

Occasionally, triple-height modules are seen (often referred to as 9U modules), which feature three connectors (P1, as well as optional P2 and P3). Triple-height modules are not specified, directly, as part of the VMEbus standard, although they often use the VMEbus standard bus system on the P1 and P2 connectors; defining the remaining pins of connector P2 and specifying some or all pins of connector P3. This format, incidentally, is used to create VXIbus systems. In such chassis formats, it is easy to understand that VMEbus modules can often be used directly in a chassis, often with little or no modification.

A chassis comprises one or (possibly) two backplanes. Connectors on backplanes are of mating types to connectors on modules: connectors numbered P1, P2, P3 (on modules) are male, while corresponding female backplane connectors are numbered J1, J2, J3. Interestingly, there appears to be no valid reason (other than some unprintable suggestions) why these letter codes for module and backplane connectors are used.

Where two backplanes are used in a chassis there is one each for the two female connector types, J1 and J2. Backplanes may hold between 2 and 21 connectors of each type, so a full specification single backplane holds 42 connectors (21 J1 connectors, 21 J2 connectors). Alternatively a minimum specification backplane holds just two J1 connectors.

This potential division of a VMEbus system into two backplanes (and hence two module sizes) is a useful facility which allows VMEbus to compete directly with simpler computer bus systems, such as those described in Chapter 6. So, although VMEbus is potentially a far more powerful system than those smaller bus systems, it is not necessarily more expensive. Indeed, as well as competing directly, in Chapter 6 it was noted how VMEbus and STEbus computer bus systems may be successfully amalgamated with VMEbus-to-STEbus couplers. On the other hand, while a single-height module chassis system allows direct competition with less powerful buses, a full specification double-height chassis system offers 32 non-multiplexed address and data bits. Further, the VMEbus Future-bus+ offers a 64-bit architecture, albeit of a multiplexed nature, which is unlikely to be surpassed at a comparable price.

Connectors are standard Eurocard connectors (IEC603.2), comprising three rows of 32-pins each on a 0.1 inch spacing, as used in the STEbus. Connector P1 pins are all assigned and are listed in Table 8.1. Table 8.2 lists pin assignments for the optional P2 connector – only the central row is defined, the other two rows are user-definable. Often, these undefined pins are used for interface connections, say, to access an internal disk drive. Connectors are located at 0.8 inch centre-to-centre intervals on a backplane.

Overall bus structure for a system using only P1 connectors is:

- 16 data lines.
- 23 address lines.
- 7 interrupt lines.
- 36 control lines.
- 6 power lines.
- 8 ground lines.

Bus structure is extended if P2 connectors are used, by a further:

- 16 data lines.
- 8 address lines.
- 3 power lines.
- 4 ground lines.
- 1 line reserved for future use.
- 64 lines for user definition.

**Table 8.1** VMEbus P1 connector pin assignments

| Pin number | Row a signal mnemonic | Row b signal mnemonic | Row c signal mnemonic |
|---|---|---|---|
| 1 | DO0 | BBSY★ | D08 |
| 2 | DO1 | BCLR★ | D09 |
| 3 | DO2 | ACFAIL★ | D10 |
| 4 | DO3 | **BGOIN★** | D11 |
| 5 | DO4 | **BG0OUT★** | D12 |
| 6 | DO5 | **BG1IN★** | D13 |
| 7 | DO6 | **BG1OUT★** | D14 |
| 8 | DO7 | **BG2IN★** | D15 |
| 9 | **GND** | **BG2OUT★** | **GND** |
| 10 | SYSCLK | **BG3IN★** | SYSFAIL★ |
| 11 | **GND** | **BG3OUT★** | BERR★ |
| 12 | DSI★ | BR0★ | SYSRESET★ |
| 13 | DSO★ | BR1★ | LWORD★ |
| 14 | WRITE★ | BR2★ | AM5 |
| 15 | **GND** | BR3★ | A23 |
| 16 | DTACK★ | AM0 | A22 |
| 17 | **GND** | AM1 | A21 |
| 18 | AS★ | AM2 | A20 |
| 19 | **GND** | AM3 | A19 |
| 20 | IACK★ | **GND** | A18 |
| 21 | **IACKIN★** | SERCLK(1) | A17 |
| 22 | **IACKOUT★** | SERDAT★(1) | A16 |
| 23 | AM4 | **GND** | A15 |
| 24 | AO7 | IRQ7★ | A14 |
| 25 | AO6 | IRQ6★ | A13 |
| 26 | AO5 | IRQ5★ | A12 |
| 27 | AO4 | IRQ4★ | A11 |
| 28 | AO3 | IRQ3★ | A10 |
| 29 | AO2 | IRQ2★ | AO9 |
| 30 | AO1 | IRQ1★ | AO8 |
| 31 | −12 V | **+5VSTDBY** | +12 V |
| 32 | +5 V | +5 V | +5 V |

★ Designates an active low signal

*Note:* Signal assignments shown in bold print in above table **are not** terminated. All remaining 72 signal lines **are** terminated.

**Table 8.2**    VMEbus P2 connector pin assignments

| Pin number | Row a signal mnemonic | Row b signal mnemonic | Row c signal mnemonic |
|---|---|---|---|
| 1 | User defined | **+5 V** | User defined |
| 2 | User defined | **GND** | User defined |
| 3 | User defined | RESERVED | User defined |
| 4 | User defined | A24 | User defined |
| 5 | User defined | A25 | User defined |
| 6 | User defined | A26 | User defined |
| 7 | User defined | A27 | User defined |
| 8 | User defined | A28 | User defined |
| 9 | User defined | A29 | User defined |
| 10 | User defined | A30 | User defined |
| 11 | User defined | A31 | User defined |
| 12 | User defined | **GND** | User defined |
| 13 | User defined | **+5 V** | User defined |
| 14 | User defined | D16 | User defined |
| 15 | User defined | D17 | User defined |
| 16 | User defined | D18 | User defined |
| 17 | User defined | D19 | User defined |
| 18 | User defined | D20 | User defined |
| 19 | User defined | D21 | User defined |
| 20 | User defined | D22 | User defined |
| 21 | User defined | D23 | User defined |
| 22 | User defined | **GND** | User defined |
| 23 | User defined | D24 | User defined |
| 24 | User defined | D25 | User defined |
| 25 | User defined | D26 | User defined |
| 26 | User defined | D27 | User defined |
| 27 | User defined | D28 | User defined |
| 28 | User defined | D29 | User defined |
| 29 | User defined | D30 | User defined |
| 30 | User defined | D31 | User defined |
| 31 | User defined | **GND** | User defined |
| 32 | User defined | **+5 V** | User defined |

*Note:* Signal assignments shown in bold print in table above are **not** terminated. All remaining 25 signal lines **are** terminated.

## Functional modules and data transfer bus cycles

To help the understanding of VMEbus and its associated parts, various tasks and system signals are allocated to **functional modules**. These functional modules are, however, purely conceptual and designed as aids to understanding rather than aids to design – actual modules plugged into a VMEbus chassis need not physically follow functional modules' logical specifications.

Crucial to the understanding of functional modules and VMEbus operation is the fact that important processes occur in predefined **cycles**. One example is data transfer, which occurs in processes known as **data transfer bus cycles**.

Equally crucial is the fact that VMEbus lines are divided up, purely for ease of use, into four **sub-buses** each having a specific function:

- **Data transfer bus** (DTB) – used for transfer of binary data between VMEbus modules. Comprises address lines, data lines and control signals.
- **DTB arbitration bus** – used to ensure only one module controls the data transfer bus at one time. Comprises bus request lines, bus grant lines, bus busy and clear lines.
- **Priority interrupt bus** – gives up to seven levels of interrupt. Comprises interrupt request lines and interrupt acknowledge lines.
- **Utility** bus – used to maintain system operation, reset and timing.

These sub-buses are shown in Figure 8.1, where available functional modules are also shown. VMEbus signals existing on the four sub-buses are listed in Table 8.3, as mnemonics under sub-bus name. Note those signal mnemonics marked ★ indicate signals which are true when the signal is low, or whose actions are initiated on a high to low transition.

Functional modules used in VMEbus are:

- **Master** – a module able to initiate data transfer bus cycles over VMEbus.
- **Slave** – a module which is able to detect data transfer bus cycles generated by masters, and participate if selected.
- **Bus timer** – a module which times data transfer bus cycles, terminating if too much time is taken thereby preventing potential lockups.
- **Interrupt handler** – when an interrupt is detected, this module takes required action (initially gaining control of the data transfer bus – see later – then generating an interrupt acknowledge cycle to detect which module has requested an interrupt).
- **Interrupter** – a module which requests an interrupt from an interrupt handler (also known as an **interrupt requester**).

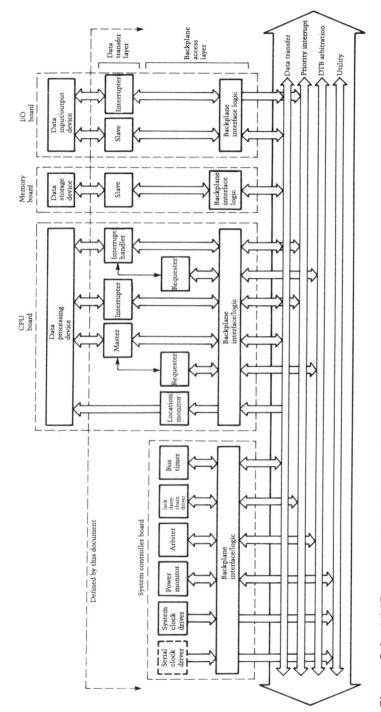

**Figure 8.1**  *VMEbus as a block diagram, showing sub-buses and functional modules. Also shown are protocol layers: backplane access layer and data transfer layer*

**Table 8.3** VMEbus signals as mnemonics under sub-buses

| Sub-bus | Lines |
|---|---|
| Data transfer bus | A01 to A31, AM0 to AM5, DS0* to DS1*, LWORD*, AS*, BERR*, DTACK*, WRITE*, D00 to D31 |
| DTB arbitration bus | BR0* to BR3*, BG0IN* to BG3IN*, BG0OUT* to BG3OUT*, BBSY*, BCLR* |
| Priority interrupt bus | IRQ1* to IRQ7*, IACK*, IACKIN*, IACKOUT* |
| Utility bus | SYSCLK*, SYSRESET*, SYSFAIL*, ACFAIL* |

- **Requester** – masters and interrupt handlers use a requester (sometimes called a **bus requester**) to gain control of the data transfer bus.
- **Arbiter** – this module monitors bus interrupt requests, granting control of the data transfer bus to one module at a time.
- **JACK\* daisy chain driver** – this module polls all interrupter modules on the bus, to detect which has requested an interrupt.
- **Location monitor** – a module which monitors VMEbus, detecting when pre-programmed module addresses are selected.
- **System clock driver** – a module which generates a 16 MHz utility clock (SYSCLK) to all modules.
- **Serial clock driver** – a module which generates a clock (SERCLK) for use with the VMSbus side bus (see later). Frequency of SERCLK is dependent on VMSbus itself.
- **Power monitor** – this module monitors VMEbus power and generates system reset signals when required.

VMEbus signals generated and monitored by functional modules are listed in Table 8.4, as mnemonics under functional module name.

## Process cycles

Important processes within VMEbus are classed as cycles. There are three main processes so classed; **data transfer bus cycles**, **arbitration cycles**, and **interrupt acknowledge cycles**.

### Data transfer bus cycles

Used to allow controlled data transfer over the data transfer bus, four types of data transfer bus cycles exist:

**Table 8.4**   VMEbus signals as mnemonics generated and monitored by functional modules

| Functional module | Signals generated | Signals monitored |
|---|---|---|
| Master | AM0 to AM5<br>DS0★ to DS1★<br>IACK★<br>WRITE★<br>AS★<br>LWORD★0<br>A01 to A310<br>D00 to D310 | SYSRESET★<br>BERR★<br>DTACK★<br>ACFAIL★0<br>BCLR★0<br>D00 to D310 |
| Slave | DTACK★<br>BERR★0<br>D00 to D310 | SYSRESET★<br>AM0 to AM5★<br>IACK★<br>LWORD★<br>DS0★ to DS1★<br>WRITE★<br>AS★0<br>A01 to A310<br>D00 to D310 |
| Bus timer | | DS0★ to DS1★<br>DTACK★0<br>BERR★ 0 |
| Interrupt handler | DS0★<br>AS★<br>IACK★<br>A01 to A03<br>DS1★ 0<br>WRITE★ 0<br>LWORD★ 0 | BERR★<br>SYSRESET★<br>DTACK★<br>IRQX★ to IRQY★<br>D00 to D310 |
| Interrupter | IRQ1★ to IRQ7★<br>DTACK★<br>IACKOUT★<br>BERR★ 0<br>D00 to D310 | DS0★<br>AS★<br>IACKIN★<br>SYSRESET★<br>A01 to A03<br>DS1★ 0<br>WRITE★ 0<br>IACK★ 0<br>LWORD0 |
| Requester | BR0★ to BR3★<br>BBSY★<br>BG0OUT★ to<br>BG3OUT★ | SYSRESET★<br>BG0IN★ to BG3IN★<br><br>BR0★ to BR3★0<br>BBSY0 |

**Table 8.4**    *(cont.)*

| Functional module | Signals | |
|---|---|---|
| | *generated* | *monitored* |
| Arbiter | BG0OUT★ to | |
| | BG3OUT★ | BR0★ to BR3★ |
| | BCLR★θ | BBSY★ |
| | | SYSRESET★ |
| IACK★ daisy chain driver | IACKOUT★ | DS0★ to DS1★ |
| | | AS★ |
| | | IACKIN★ |
| | | IACK★ θ |
| Location monitor | LWORD★ | LWORD★ |
| | DS0★ to DS1★ | DS0★ to DS1★ |
| | AM0★ to AM5★ | AM0★ to AM5★ |
| | IACK★ | IACK★ |
| | WRITE★ | WRITE★ |
| | A01 to A31θ | A01 to A31φ |
| | AS★ θ | AS★ θ |
| System clock driver | SYSCLK | |
| Serial clock driver | SERCLK | |
| Power monitor | SYSRESET★ | AC power |
| | ACFAIL★ | |

*Note:* θ refers to optional signals

- **Read/write cycle** – used to transfer 8, 16, 24, or 32 bits of data between masters and slaves. Cycle begins as a master broadcasts an address and an address modifier signal. Slaves respond if selected.
- **Read-modify-write cycle** – permits indivisible bus cycles. Useful when arbitrating shared resources.
- **Block transfer cycle** – allows blocks up to 256 bytes to be transferred between masters and slaves. Cycle is quicker than multiple read/write cycles, as slaves are addressed only once.
- **Address-only cycle** – where masters generate an address, but addressed slaves do not respond prior to data transfer.

## Arbitration cycles

These occur during bus arbitration, in which modules requesting control of the data transfer bus are allocated it in order, by the arbiter functional module. The order used is predefined by the arbiter functional module in a VMEbus system, and may be:

- In strict priority, by a **prioritized arbiter** (PRI).
- In a rotating priority, by a **round–robin–select arbiter** (RRS).
- In a polled (or daisy chained) order, by a **single level arbiter** (SGL).

Initially, a requester generates a bus request (BRO★ to BR3★) signal, after which the arbiter grants the bus to the requester (BG0IN★ to BG3IN★) when the bus is not busy. The requester then asserts the bus busy signal (BBSY★) to indicate no-one else may control the bus, and negates the bus busy signal when it no longer requires the bus.

### Interrupt acknowledge cycles

Initiated by interrupt handlers in response to an interrupt, the interrupt acknowledge cycle arbitrates the interrupt sources by monitoring a STATUS/ID word received from each interrupter.

## Signals, lines and protocols

Signals within VMEbus are generally grouped into two **protocol layers**. The lowest layer, called the **backplane access layer**, comprises backplane interface logic, utility bus modules, and DTB arbitration bus modules. The higher layer is called the **transfer layer**, and comprises data transfer bus modules and priority bus modules. These two layers are shown in Figure 8.1.

Within both layers, protocols used are of two main types:

- **Closed-loop protocols**. These use **interlocked bus signals**, sent from one module (a source) to another module (a destination), usually requiring an acknowledgement. Generally, interlocked bus signals coordinate a system's internal functions, such as data flow between two modules.
- **Open-loop protocols**. These use **broadcast bus signals**, sent by one module to all other modules, to indicate special conditions. No acknowledgement is required, and the broadcast signal is merely maintained for a minimum specified time on a dedicated signal line.

Signals present in VMEbus are discussed here, as being present on lines belonging to one of the four sub-buses.

### Data transfer bus

Address, data and control lines are all present on the data transfer bus. A typical VMEbus system is illustrated in block diagram form in Figure 8.2, where functional modules using the data transfer bus are shown shaded.

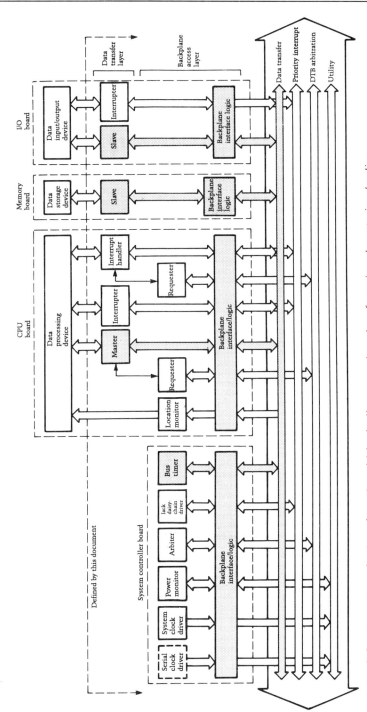

**Figure 8.2** Data transfer bus functional block diagram, showing functional parts in shading

## Address lines A01 to A31

Address lines are driven by masters and interrupt handlers. During data transfer bus cycles address lines broadcast **short** (16-bit), **standard** (24-bit), or **extended** (32-bit) addresses. Interrupt handlers also broadcast current interrupt level being acknowledged on lines A01 to A03. Address lines are three-state driven.

## Address modifier lines AM0 to AM5

These three-state lines, driven by masters and interrupt handlers, broadcast numbers of valid address lines during data transfer bus cycles, and cycle types.

## Data strobes DS0* to DS1*

Data strobes are three-state driven lines with three functions – all specified by masters and interrupt handlers. First, when used with long word line LWORD★ and address line A01, they indicate size and type of data transfer. Second, they indicate valid data during a write cycle. Third, they inform a slave it should place data on the bus during a read cycle.

## Long word LWORD*

Long word is a three-state line, driven by masters to indicate byte-location of current data transfer. Used in conjunction with address line A01, and data strobes DS0★ and DS1★.

## Data lines D00 to D31

Data lines are driven by masters, interrupters or slaves. Data lines are three-state and bidirectional. Byte-size of data (and hence number of data lines) is determined by data strobes DS0★ and DS1★, in conjunction with address line A01 and long word LWORD★.

## Address strobe AS*

This is a high-current three-state line, driven by masters and interrupt handlers, which indicates valid address and address modifier on the address bus.

## Bus error BERR*

This is an open-collector line, driven by an addressed slave or bus timer to indicate incorrect data transfer.

## Data transfer acknowledge DTACK*

This is an open-collector line, driven by an addressed slave or interrupter. The falling edge indicates valid data on the data bus during a read cycle, or data has been accepted during a write cycle.

## Write WRITE*

This is a three-state signal, driven by a master. High indicates a read cycle; low indicates a write cycle.

## DTB arbitration bus

The DTB arbitration bus comprises only control lines, with the purpose of allowing VMEbus to be used by several processors. A typical VMEbus system is illustrated in block diagram form in Figure 8.3, where functional modules using the DTB arbitration bus are shown shaded.

### Bus request BR0* to BR3*

These are open-collector lines, driven by requesters to indicate a master requires use of the data transfer bus.

### Bus grant in BG0IN* to BG3IN*

These are totem-pole lines, driven by the arbiter. A received *bus grant in* signal indicates the receiving board is granted use of the data transfer bus. These, together with *bus grant out* lines form bus grant daisy-chains.

### Bus grant out BG0OUT* to BG3OUT*

These are totem-pole lines, used in a bus grant daisy-chain, which indicate to the next module in the daisy-chain it is granted use of the data transfer bus.

### Bus busy BBSY*

An open-collector line, used by a requester to indicate to the arbiter its master is using the data transfer bus.

### Bus clear BCLR*

This is a high-current totem-pole line, driven by the arbiter and indicating to the current master a request for the data transfer bus.

## Priority interrupt bus

This bus comprises seven interrupt request lines, together with three control lines which are used to select which module is to be serviced. A typical VMEbus system is illustrated in block diagram form in Figure 8.4, where functional modules using the priority interrupt bus are shown shaded.

### Interrupt request IRQ1* to IRQ7*

These are open-collector lines, driven by interrupters to request an interrupt. Higher numbered lines have higher priorities.

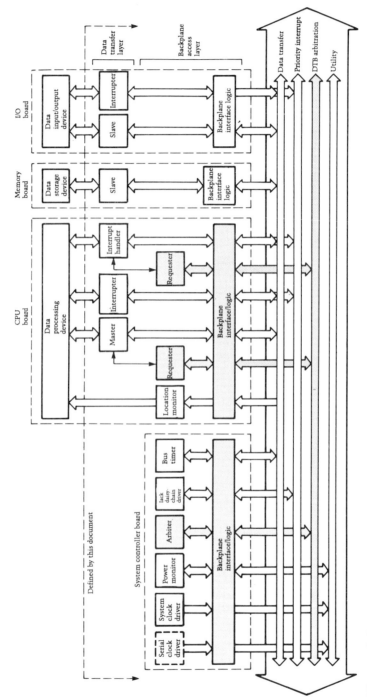

**Figure 8.3**   *DTB arbitration bus functional block diagram, showing functional parts in shading*

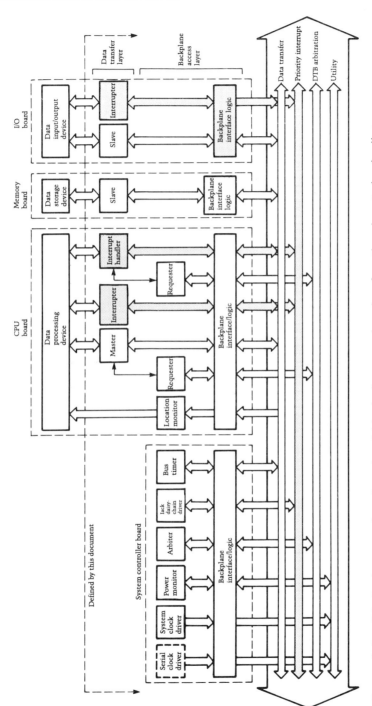

**Figure 8.4** *Priority interrupt bus functional block diagram, showing functional parts in shading*

### Interrupt acknowledge IACK*

An open-collector or a three-state line driven by an interrupt handler to acknowledge an interrupt request. It is routed via the backplane to the IACKIN* pin of slot 1 module, where it is monitored by the IACK daisy-chain driver.

### Interrupt acknowledge in IACKIN*

A daisy-chained totem-pole line, received from the previous module in the daisy-chain, which indicates to the receiving module that it is allowed to respond to the interrupt acknowledge cycle in progress. Previous module's IACKOUT* line forms the following module's IACKIN* line.

### Interrupt acknowledge out IACKOUT*

A daisy-chained totem-pole line, sent to the next module in the daisy-chain, indicating that the next module may respond to the current interrupt acknowledge cycle. IACKOUT* line of one module forms the following module's IACKIN* line.

## Utility bus

Comprising only four lines, the utility bus provides facilities for system initialization, timing and control in the event of power failure. A typical VMEbus system is illustrated in block diagram form in Figure 8.5, where functional modules using the utility bus are shown shaded.

### System clock SYSCLK*

This is a high-current totem-pole line driven by slot 1 system controller. It is a 16 MHz signal for general-purpose use, but is not used for specific bus timing purposes as VMEbus is asynchronous.

### System reset SYSRESET*

An open-collector line, driven by any module, which indicates a system reset is in progress.

### System fail SYSFAIL*

An open-collector line, driven by any module, which indicates a failure has occurred in the system. Specific failures and responses are defined by the user.

### AC failure ACFAIL*

This is an open-collector line, driven by the power monitor module, which indicates to all modules the power source is about to disappear.

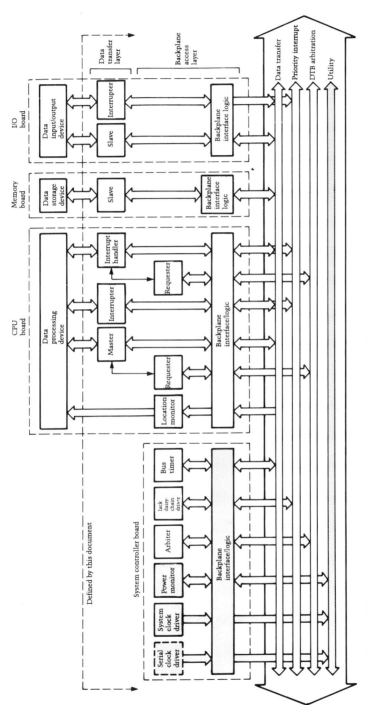

**Figure 8.5** *Utility bus functional block diagram, showing functional parts in shading*

### Other lines

In addition to lines on specific buses within VMEbus, there are a number of other lines which, although present on VMEbus backplanes, are not classed as forming a bus of their own.

*Ground GND*
DC voltage signal reference and power return path.

*+5V, +12V, −12V*
DC system power supplies.

*+5V STDBY*
Standby power supply for modules which require battery backup in the case of power failure.

*Serial clock SERCLK*
This is a totem-pole line, used for timing serial data transmissions using VMSbus (see later).

*Serial data SERDAT*
This is an open–collector line, used for serial data transmissions using VMSbus (see later).

*Reserved RESERVED*
A signal line reserved for future use.

## Electrical considerations

Most electrical specifications of VMEbus may be conveniently grouped into a number of rules, the main ones of which are:

- Signal conductors on backplanes have a maximum length of 500 mm (19.68 inch).
- No more than 21 slots may exist in a backplane.
- Most signal lines are terminated at both ends of the backplane. Table 8.5 lists all signal lines as mnemonics, signal names, driver types and whether bused or terminated.
- No steady state signal on the backplane is more than +5 V or less than 0 V.
- Steady state driver low output level is ≤0.6 V, while driver high output voltage is ≥2.4 V.
- Steady state receiver low input level is ≤0.8 V, while receiver high input level is ≥2.0 V.

**Table 8.5** VMEbus signal lines as mnemonics, with names, driver types, and whether bused and terminated or not

| Mnemonic | Name | Driver type | Bused and terminated |
|---|---|---|---|
| A01 to A31 | address lines | standard three-state | yes |
| ACFAIL* | AC power failure | open collector | yes |
| AM0 to AM5 | address modifier | standard three-state | yes |
| AS* | address strobe | high-current three-state | yes |
| BBSY* | bus busy | open collector | yes |
| BCLR* | bus clear | high-current totem-pole | yes |
| BERR* | bus error | open collector | yes |
| BG0IN* to BG3IN* | bus grant daisy-chain | standard totem-pole | no |
| BG0OUT* to BG3OUT* | | | |
| BR0* to BR3* | bus request | open collector | yes |
| DS0* to DS1* | data strobes | high-current three-state | yes |
| DTACK* | data transfer acknowledge | open collector | yes |
| IACK* | interrupt acknowledge | standard three-state | yes |
| IACKIN*/IACKOUT* | interrupt acknowledge daisy-chain | standard totem-pole | no |
| IRQ1* to IRQ7* | interrupt request | open collector | yes |
| LWORD* | long word | standard three-state | yes |
| RESERVED | reserved | — | yes |
| SERCLK | serial clock | high-current totem-pole | yes |
| SERDAT* | serial data | open collector | yes |
| SYSCLK | system clock | high-current totem-pole | yes |
| SYSFAIL* | system failure | open collector | yes |
| SYSRESET* | system reset | open collector | yes |
| WRITE* | write | standard three-state | yes |
| +5 V, +12 V, −12 V | power supplies | — | bused, not terminated |
| GND | DC signal reference and power return | — | bused, not terminated |

- All modules provide clamping on each used signal line, to ensure negative excursions below −1.5 V do not occur.
- Printed circuit board track length from connectors to circuit components does not exceed 50.8 mm (2 inch).

## Side buses

Some applications of VMEbus require data transmission at a faster rate than VMEbus itself allows. Here a secondary data channel may be used, called a **side bus**, which allows modules to communicate directly via a private backplane. Figure 8.6 illustrates this. A number of side buses exist: VMXbus, VSBbus, VMSbus are three examples, and are usually chosen for specific attributes such as high speed (VMXbus), more efficient interrupt cycling (VSBbus) or synchronous serial data transmission (VMSbus).

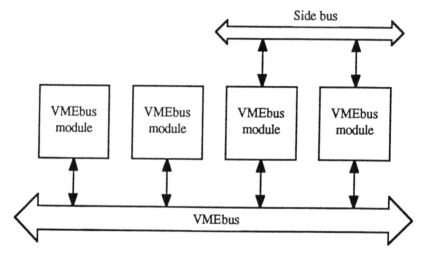

**Figure 8.6**  *Showing VMEbus with modules, together with a side-bus allowing direct module-to-module communication*

In the respects defined here for side buses, there is another side bus of VMEbus which is used to extend basic VMEbus performance. VXIbus (which is a contraction of *VMEbus extensions for instrumentation*) is a bus system which uses most of the electrical, mechanical and logical features of VMEbus, adding to it a number of other buses and facilities, tailoring it specifically for a computer bus for use in test and instrumentation systems, particularly automatic test equipment systems. It is, of course, of such importance to automatic test equipment systems it warrants its own description – given in the following chapter.

## Futurebus+ extended architecture

A revision to the VMEbus standard IEEE1014 is currently proposed. Revision D which is, at the time of writing, at draft stage looks to extend VMEbus to allow 64-bit address and data facilities. The new specification is currently called **Futurebus+ extended architecture**. Although information is merely preliminary at present, it is possible to outline major changes proposed to the current VMEbus standard to describe how Futurebus+ works.

To allow 64-bit address and data facilities, address and data lines (of 32-bits each) are used together in a multiplexed, bidirectional way to create a word length of 64 bits. Thus addressing is undertaken with address lines A01 to A31 and D00 to D31 in parallel which, together with address modifiers AM0 to AM5, long word line LWORD*, and data strobes DS0* and DS1* allows up to 64-bit wide addresses (known as **long addresses**) to be accessed.

Similarly data transfer up to 64-bits wide is possible, using data and address lines in parallel together with long word line LWORD*.

An extra control: retry line RETRY* is included in the data transfer bus to notify the controlling master if a data transfer is not completed. Master will retry the transfer at a later time.

## Further reading

Readers wishing to develop further their understanding of VMEbus are referred to:

- *IEEE1014-1987*, or *IEC821* and *IEC297*: the *VMEbus specification*.
- *The VMEbus Handbook*, by Wade D. Peterson.
- *IEEE P1014 revision D draft*: the VME/Futurebus+ extended architecture draft revision D proposal for the VMEbus specification.

from which most information for this chapter was taken. All three are available from the VMEbus International Trade Association.

Further information is given in *The VMEbus User's Handbook* by Steve Heath published by Butterworth-Heinemann, Oxford.

# 9 VXIbus

One of the most important developments to occur in automatic test equipment to date, VXIbus is a successful attempt to take out some of the mystery associated with automatic test procedures and systems. Previously, automatic test equipment had been the province of either large, expensive turnkey test systems or rack-and-stack GPIB-based on test systems of whatever classification – component test, continuity test, in-circuit test, functional test and so on. Of necessity these systems are specific to one measurement application, and adaptable to other applications only with considerable hardware and software changes, consequently involving considerable expense.

Since 1981, though, automatic test equipment solutions had been constructed – particularly under requests from the United States Department of Defense – using the VMEbus computer system bus. VMEbus automatic test equipment systems afford a considerable reduction in size compared with corresponding turnkey solutions, and a well-designed backplaned bus ensures far higher data rates than rack-and-stack systems can ever achieve. The result was (and still is, for VMEbus systems are still popular) smaller, faster automatic test equipment systems. There are, in fact, thousands of modular cards available for VMEbus systems, from hundreds of manufacturers. VMEbus still has its limitations, though, based mainly on the fact the VMEbus standard is essentially just a general-purpose computer bus standard, and does not specify system details. Indeed, most of the VMEbus modular cards available are computer-orientated; only a few are available which can be remotely considered suitable for automatic test equipment systems (generally data acquisition modules for industrial process and control).

In 1987 five large test instrument and equipment manufacturers: Colorado Data Systems; Hewlett Packard; Racal Dana Instruments; Tektronix; Wavetek, formed a consortium to devise the additional specifications required to make an automatic test equipment *system* standard from VMEbus. The consortium's aim was a complete standard utilizing a VMEbus-based backplane bus, Eurocard standard dimensions, and instrumentation facilities and features present in the GPIB. The result is VXIbus (in fact, VXIbus stands for *VME extensions for instrumentation bus*).

Having defined the VXIbus specification, the five manufacturers set about developing test equipment instruments to use in a VXIbus automatic test equipment system. Many other companies have joined (Bruel & Kjaer, John Fluke, GenRad, Keithley Instruments, National Instruments and so on) the consortium. Other companies provide VXIbus products such as backplanes, bus testers, specific module instruments and software. In less than just four years since conception VXIbus appears to be set to revolutionize manufacture of automatic test equipment systems.

## Basic VXIbus system components

As it is a computer system specification, albeit geared for use in automatic test equipment systems, VXIbus has a number of areas common to general-purpose computer systems, which are considered here as part of an initial consideration of the specification.

### *Hardware*

The physical structure of a VXIbus automatic test equipment system comprises a mainframe chassis with backplane and connections to allow modules to be plugged in as required (Figure 9.1). It is thus a classical modular-based backplaned computer system. However, logical design is structured specifically with an automatic test equipment system in mind. Consequently, a VXIbus automatic test equipment system has the benefits of high data rate associated with a modular-based backplaned computer system, coupled with the ease-of-use and controllability of a purpose-built automatic test equipment system. Modules are inserted into slots in the backplaned housing. Slot 0 is always occupied by a timing and control module.

An individual test instrument for use on VXIbus is known somewhat ambiguously as a **device**. A single module on VXIbus may house one or more devices. On the other hand, a complex device may be formed by one or more modules. Both these extremes are catered for by the logical structure of VXIbus. As modules are often referred to as **cards** and these cards may hold a complete test instrument, the term **instrument-on-a-card** (IAC) is commonly used to describe the modular basis of VXIbus devices.

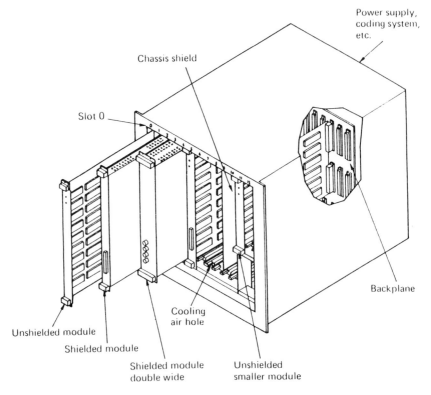

**Figure 9.1**   *Classical, modular-based backplaned bus system used in VXIbus*

### Subsystems

Up to 13 modules (one of which is a timing module) may be housed in a single housing called a **subsystem**. If a larger system is required, individual subsystems may be connected, to a maximum of 256 devices. To connect subsystems a number of interface methods may be used, including IEC625 (IEEE488 – the general purpose interface bus), and V24 (EIA232 – the old R S232).

The VXIbus subsystem is important from an electrical point of view as there are significant local signals specific to the subsystem on the backplane. Examples are ECL clocks and trigger signals described later. These local signals cannot be directly connected to other subsystems, but must instead be interfaced with suitable buffering interface. Further, VXIbus subsystem backplanes feature high performance ECL signals with maximum timing delays of 5 ns and timing skewing within 2 ns, supporting extremely close time coordination between instrument modules within a subsystem.

**Photo 9.1** *Mainframe VXIbus chassis, complete with backplaned connections (Tektronix)*

**Photo 9.2** *Universal counter/timer module for use on VXIbus subsystems (Tektronix)*

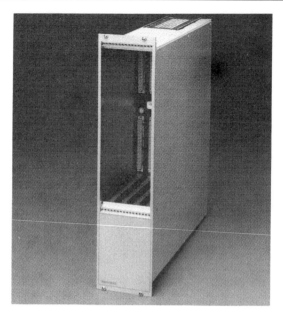

**Photo 9.3**   *VME-VXI conversion module for a VXIbus system (Tektronix)*

Consequently, such close time coordination *is* only possible *within* a subsystem and cannot be maintained over interfaces between subsystems.

The logical structure of VXIbus is totally open. That is, operating system, microprocessor, computer interface, hierarchy and so on are all undefined. Only protocols and criteria which must exist to ensure compatibility between modules are defined, and these do not interfere with individual device microprocessor controllers.

VXIbus hardware structure is based fairly closely on the VMEbus, incorporating the complete VMEbus backplaned structure using P1 and P2 connectors between modules and backplane, and A- and B-sized modules, adding to it in a number of ways.

### VXIbus comparison with VMEbus

A minimum specification VMEbus system requires only the P1 connector. Handshaking, arbitration and interrupt support is present on connector P1, which supports 16 or 24 address bits and 8 or 16 data bits. Connector P2 is used to expand a VMEbus system to 32 address bits and 32 data bits. Extra lines required are carried on the central row of pins on connector P2, while outer rows are undefined – for user-specification. These undefined outer rows are typically used in VMEbus systems for interface connections

**Table 9.1** VXIbus module sizes and connectors

| Module | Connectors | Sizes (mm) |
|--------|-----------|-----------|
| A | P1 | 99 by 160 by 20 |
| B | P1 | 233 by 160 by 20 |
|   | P2 (optional) | |
| C | P1 | 233 by 340 by 30 |
|   | P2 (optional) | |
| D | P1 | 365 by 340 by 30 |
|   | P2, P3 (optional) | |

between modules and, say, an internal disk drive, chassis-mounted connectors, or other modules. For example, VSB (VMEbus subsystem bus) is a commonly used bus which defines the outer pin rows of the P2 connector specifically for use as a communication bus between up to six VMEbus modules. Connectors used in VXIbus are, though, identical physically to those of VMEbus, and are defined by IEC603.2.

Summarizing, the basic differences VXIbus features over VMEbus include:

- Slightly wider spacing (30 mm against 25 mm VMEbus modular slot spacing) between slots of modules. This allows VXIbus modules to hold large analog components or daughter boards, and allows a manufacturer to shield modules in metal cases. On the other hand, wider slot spacing means VMEbus modules can fit into a VXIbus chassis, but not vice versa.
- A complete specification of packaging requirements for modules; electromagnetic compatibility, power distribution, cooling and air flow in the chassis, and so on.
- Two new module sizes, C- and D-sized modules, with a further optional connector P3 on the largest D-sized module. Table 9.1 lists all VXIbus module sizes and connectors.
- Definition of remaining pin assignments on the P2 connector, and all P3 connector assignments.
- Possibilities of low level and high level communications between devices on the bus.

## Electrical architecture

Electrical connections between modules in a VXIbus subsystem are all made through the backplane. Connections are grouped into different buses, and each bus is used for specific purposes in a particular system; some systems use all buses, others use only some buses.

**Table 9.2**   VXIbus system buses and types

| Bus | Type |
|---|---|
| VMEbus computer bus | global |
| Trigger bus | global |
| Analog sumbus | global |
| Power distribution bus | global |
| Clock and synchronization bus | unique |
| Star bus | unique |
| Module identification bus | unique |
| Local bus | private |

These buses are logically grouped into a total of eight (although a few reserved pins exist, too) and these are listed in Table 9.2 along with a description of bus type: global (always accessible and shared by all modules); unique (from slot 0 timing and control module to other modules on a one-to-one basis); private (local buses between logically adjacent modules).

VXIbus standard requires only connector P1 to be used in the minimum configuration between a module and the backplane. Thus, all global buses listed in Table 9.2 must be present (at least in a minimum specification – extra VMEbus data and address bits are carried on the P2 connector) on that connector. Consequently connector P1 is the VXIbus backbone, which provides all signals and voltages required to construct a computer bus-based automatic test equipment system. A P1-alone system can have an 8- or 16-bit data bus, with a 64 Kb or 16 Mb address space.

If modules have one or both other connectors, system performance can be increased radically. For example, connector P2 has the extra VMEbus address and data bits to allow a computer-based system of 4 Gb address space to be constructed. Trigger, analog sumbus, clock and synchroniz-ation, module identification, and local buses listed in Table 9.2 are also included in connector P2, while connector P3 includes the star bus and further trigger, power distribution, clock and synchronization, and local bus lines.

The overall electrical structure of a VXIbus system with only P1 connector is identical to that of VMEbus with only P1 connector. Readers should refer to Chapter 8 for a description of VMEbus.

The overall structure of a VXIbus system with P2 connector is:

- The VMEbus.
- 13 module identity (MODID) lines.
- 12 local bus (LBUS) lines.
- 8 TTL trigger (TTLTR) lines.

- 2 ECL trigger (ECLTR) lines.
- 1 analog summing bus (SUMBUS) lines.
- 2 clock (CLK10) lines – at 10 MHz.
- 2 lines reserved for future use.

The overall structure of a VXIbus system with P3 connector is:

- VXIbus system with P2 connectors.
- 24 additional local bus lines.
- 4 additional ECL trigger lines.
- 52 star trigger (STAR) lines for precision module-to-module timing.
- 2 clock (CLK100) lines – at 100MHz (synchronous with P2 CLK10).
- 2 synchronizing signal (SYNC100) lines.
- 2 additional power supply lines.
- 4 lines reserved for future use.

Table 9.3 lists pin assignments for the fully defined P2 connector in a slot 0, while Table 9.4 lists assignments for other slots. Similarly, Table 9.5 lists pin assignments for the P3 connector in slot 0, while Table 9.6 lists assignments for other slots.

Electrical architecture of a possible VXIbus system is illustrated in Figure 9.2, showing graphically the three different types of buses of the VXIbus listed in Table 9.2.

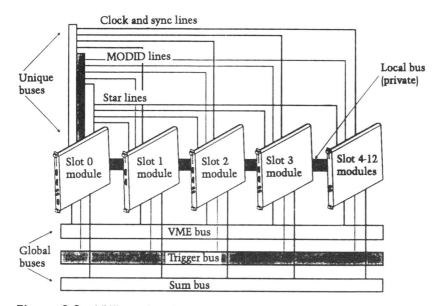

**Figure 9.2** *VXIbus electrical architecture, showing relationships of global, unique and private buses*

**Table 9.3**    VXIbus slot 0, P2 connector pin assignments

| Pin Number | Row a signal mnemonic | Row b signal mnemonic | Row c signal mnemonic | Pin number |
|---|---|---|---|---|
| 1 | ECLTRG0 | +5 V | CLK10+ | 1 |
| 2 | −2 V | GND | CLK10− | 2 |
| 3 | ECLT RG1 | RSV1 | GND | 3 |
| 4 | GND | A24 | −5.2 V | 4 |
| 5 | MODID12 | A25 | LEBUSC00 | 5 |
| 6 | MODID11 | A26 | LBUSC01 | 6 |
| 7 | −5.2 V | A27 | GND | 7 |
| 8 | MODID10 | A28 | LBUSC02 | 8 |
| 9 | MODID09 | A29 | LBUSC03 | 9 |
| 10 | GND | A30 | GND | 10 |
| 11 | MODID08 | A31 | LBUSC04 | 11 |
| 12 | MODID07 | GND | LBUSC05 | 12 |
| 13 | −5 2 V | +5 V | −2 V | 13 |
| 14 | MJODID06 | D16 | LBUSC06 | 14 |
| 15 | MODID05 | D17 | LBUSC07 | 15 |
| 16 | GND | D18 | GND | 16 |
| 17 | MODID04 | D19 | LBUSC08 | 17 |
| 18 | MODID03 | D20 | LBUSC09 | 18 |
| 19 | −5.2 V | D21 | −5.2 V | 19 |
| 20 | MODID02 | D22 | LBUSC10 | 20 |
| 21 | MODID01 | D23 | LBUSC11 | 21 |
| 22 | GND | GND | GND | 22 |
| 23 | TTLTRG0★ | D24 | TTLTRG1★ | 23 |
| 24 | TTLTRG2★ | D25 | TTLTRG3★ | 23 |
| 25 | +5 V | D26 | GND | 25 |
| 26 | TTLTRG4★ | D27 | TTLTRG5★ | 26 |
| 27 | TTLTRG6★ | D28 | TTLTRG7★ | 27 |
| 28 | GND | D29 | GND | 28 |
| 29 | RSV2 | D30 | RSV3 | 29 |
| 30 | MODID00 | D31 | GND | 30 |
| 31 | GND | GND | =24 V | 31 |
| 32 | SUMBUS | +5 V | −24 V | 32 |

**Table 9.4** VXIbus slots 1 to 12, P2 connector pin assignments

| Pin Number | Row a signal mnemonic | Row b signal mnemonic | Row c signal mnemonic | Pin number |
|---|---|---|---|---|
| 1 | ECLTRG0 | +5 V | CLK10+ | 1 |
| 2 | −2 V | GND | CLK10− | 2 |
| 3 | ECLTRG1 | RSV1 | GND | 3 |
| 4 | GND | A24 | −5.2 V | 4 |
| 5 | LBUSA00 | A25 | LEBUSC00 | 5 |
| 6 | LBUSA01 | A26 | LBUSC01 | 6 |
| 7 | −5.2 V | A27 | GND | 7 |
| 8 | LBUSA02 | A28 | LBUSC02 | 8 |
| 9 | LBUSA03 | A29 | LBUSC03 | 9 |
| 10 | GND | A30 | GND | 10 |
| 11 | LBUSA04 | A31 | LBUSC04 | 11 |
| 12 | LBUSA05 | GND | LBUSC05 | 12 |
| 13 | −5.2 V | +5 V | −2 V | 13 |
| 14 | LBUSA06 | D16 | LBUSC06 | 14 |
| 15 | LBUSA07 | D17 | LBUSC07 | 15 |
| 16 | GND | D18 | GND | 16 |
| 17 | LBUSA08 | D19 | LBUSC08 | 17 |
| 18 | LBUSA09 | D20 | LBUSC09 | 18 |
| 19 | −5.2 V | D21 | −5.2 V | 19 |
| 20 | LBUSA10 | D22 | LBUSC10 | 20 |
| 21 | LBUSA11 | D23 | LBUSC11 | 21 |
| 22 | GND | GND | GND | 22 |
| 23 | TTLTRG0★ | D24 | TTLTRG1★ | 23 |
| 24 | TTLTRG2★ | D25 | TTLTRG3★ | 24 |
| 25 | +5 V | D26 | GND | 25 |
| 26 | TTLTRG4★ | D27 | TTLTRG5★ | 26 |
| 27 | TTLTRG6★ | D28 | TTLTRG7★ | 27 |
| 28 | GND | D29 | GND | 28 |
| 29 | RSV2 | D30 | RSV3 | 29 |
| 30 | MODID | D31 | GND | 30 |
| 31 | GND | GND | +24 V | 31 |
| 32 | SUMBUS | +5 V | −24 V | 32 |

**Table 9.5**   VXIbus slot 0, P3 connector pin assignments

| Pin Number | Row a signal mnemonic | Row b signal mnemonic | Row c signal mnemonic | Pin number |
|---|---|---|---|---|
| 1 | ECLTRG2 | +24 V | +12 V | 1 |
| 2 | GND | −24 V | −12 V | 2 |
| 3 | ECLTRG3 | GND | RSV4 | 3 |
| 4 | −2 V | RSV5 | +5 V | 4 |
| 5 | ECLTRG4 | −5.2 V | RSV6 | 5 |
| 6 | GND | RSV7 | GND | 6 |
| 7 | ECLTRG5 | +5 V | −5.2 V | 7 |
| 8 | −2 V | GND | GND | 8 |
| 9 | STARY12+ | +5 V | STARX01+ | 9 |
| 10 | STARY12− | STARY01− | STARX01− | 10 |
| 11 | STARX12+ | STARX12− | STARY01+ | 11 |
| 12 | STARY11+ | GND | STARX02+ | 12 |
| 13 | STARY11− | STARY02− | STARX02− | 13 |
| 14 | STARX11+ | STARX11− | STARY02+ | 14 |
| 15 | STARY10+ | +5 V | STARX03+ | 15 |
| 16 | STARY10− | STARY03− | STARX03− | 16 |
| 17 | STARX10+ | STARX10− | STARY03+ | 17 |
| 18 | STARY09+ | −2 V | STARX04+ | 18 |
| 19 | STARY09− | STARY04− | STARX04− | 19 |
| 20 | STARX09+ | STARX09− | STARY04+ | 20 |
| 21 | STARY08+ | GND | STARX05+ | 21 |
| 22 | STARY08− | STARY05− | STARX05− | 22 |
| 23 | STARX08+ | STARX08− | STARY05+ | 23 |
| 24 | STARY07+ | +5 V | STARX06+ | 24 |
| 25 | STARY07− | STARY06− | STARX06− | 25 |
| 26 | STARX07+ | STARX07− | STARY06+ | 26 |
| 27 | GND | GND | GND | 27 |
| 28 | STARX+ | −5.2 V | STARY+ | 28 |
| 29 | STARX− | GND | STARY− | 29 |
| 30 | GND | −5.2 V | −5.2 V | 30 |
| 31 | CLK100+ | −2 V | SYNC100+ | 31 |
| 32 | CLK100 | GND | SYNC100− | 32 |

**Table 9.6** VXIbus slot 1 to 12, P3 connector pin assignments

| Pin Number | Row a signal mnemonic | Row b signal mnemonic | Row c signal mnemonic | Pin number |
|---|---|---|---|---|
| 1 | ECLTRG2 | +24 V | +12 V | 1 |
| 2 | GND | −24 V | −12 V | 2 |
| 3 | ECLTRG3 | GND | RSV4 | 3 |
| 4 | −2 V | RSV5 | +5 V | 4 |
| 5 | ECLTRG4 | −5.2 V | RSV6 | 5 |
| 6 | GND | RSV7 | GND | 6 |
| 7 | ECLTRG5 | +5 V | −5.2 V | 7 |
| 8 | −2 V | GND | GND | 8 |
| 9 | LBUSA12 | +5 V | LBUSC12 | 9 |
| 10 | LBUSA13 | LBUSC15 | LBUSC13 | 10 |
| 11 | LBUSA14 | LBUSA15 | LBUSC14 | 11 |
| 12 | LBUSA16 | GND | LBUSC16 | 12 |
| 13 | LBUSA17 | LBUSC19 | LBUSC17 | 13 |
| 14 | LBUSA18 | LBUSA19 | LBUSC18 | 14 |
| 15 | LBUSA20 | +5 V | LBUSC20 | 15 |
| 16 | LBUSA21 | LBUSC23 | LBUSC21 | 16 |
| 17 | LBUSA22 | LBUSA23 | LBUSC22 | 17 |
| 18 | LBUSA24 | −2 V | LBUSC24 | 18 |
| 19 | LBUSA25 | LBUSC27 | LBUSC25 | 19 |
| 20 | LBUSA26 | LBUSA27 | LBUSC26 | 20 |
| 21 | LBUSA28 | GND | LBUSC28 | 21 |
| 22 | LBUSA29 | LBUSC31 | LBUSC29 | 22 |
| 23 | LBUSA30 | LBUSA31 | LBUSC30 | 23 |
| 24 | LBUSA32 | +5 V | LBUSC32 | 24 |
| 25 | LBUSA33 | LBUSC35 | LBUSC33 | 25 |
| 26 | LBUSA34 | LBUSA35 | LBUSC34 | 26 |
| 27 | GND | GND | GND | 27 |
| 28 | STARX+ | −5.2 V | STARY+ | 28 |
| 29 | STARX− | GND | STARY− | 29 |
| 30 | GND | −5.2 V | −5.2 V | 30 |
| 31 | CLK100+ | −2 V | SYNC100+ | 31 |
| 32 | CLK100− | GND | SYNC100− | 32 |

## System buses

VXIbus system buses are:

- VMEbus computer bus.
- Trigger bus.
- Analog sumbus.
- Power distribution bus.
- Clock and synchronization bus.
- Star bus.
- Module identification bus.
- Local bus.

### VMEbus computer bus

Comprising four individual buses, the VMEbus computer bus features all primary requirements of a complete computer system bus, which are:

- Data transfer bus.
- Data transfer bus arbitration bus.
- Priority interrupt bus.
- Utilities bus.

As these are covered more specifically in Chapter 8, readers are referred there for further details.

### Trigger bus

This bus is used primarily for intermodule communications purposes. Any module, including that located in slot 0 (timing and control), may use this bus to send or receive information to or from other modules. Trigger lines on the bus are general-purpose logic lines, usable for triggering, hand-shaking, clocking or data transmission purposes.

VXIbus trigger bus comprises eight transistor-transistor logic trigger lines (known as TTLTRG★ lines) and six emitter-coupled logic trigger lines (ECLTRG). All TTLTRG★ lines and two ECLTRG lines are located on connector P2, while the four remaining ECLTRG lines are on connector P3. The primary operational difference between the two types of lines is operational speed: maximum clock transmission for TTLTRG★ lines is 12.5 MHz, while ECLTRG lines' maximum clock transmission is 62.5 MHz; although other differences exist. Figure 9.3 shows the basic layout of TTLTRG★ and ECLTRG trigger lines of a VXIbus system.

### TTLTRG★ lines

TTLTRG★ lines are open collector TTL lines, bused over the whole length of a VXIbus subsystem backplane. Lines are terminated with pull-up

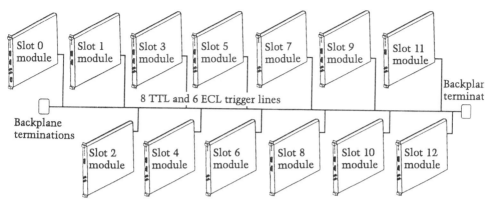

**Figure 9.3**   *Basic layout of VXIbus TTLTRG\* and ECLTRG trigger lines*

terminations, which provide constant non-asserted high states on individual lines unless connected modules assert low states. Use of backplaned pull-up terminations means rise times of these lines (functions of passive pull-up terminations) are considerably greater than fall times (which are functions of active TTL drivers). Consequently, it is usual to use falling edges of transmissions for timing purposes, not longer rising edges. Further, due also to the difference in rise times and fall times, a number of protocols exist which define separate timing requirements for trigger sources and trigger acceptors.

TTLTRG★ lines follow some basic rules:

- Lines are terminated by backplaned $100\,\Omega$ impedance terminations (often $50\,\Omega$ impedance terminations are used, without significantly affecting performance).
- A module's interface to TTLTRG★ lines conforms to standard driving and loading rules for open collector lines.
- All lines become unasserted within one second of a required reset.
- Modules allocate TTLTRG★ lines to specific functions in groups of one, two or four.
- If single-line TTLTRG★ groups are used, modules connect to any TTLTRG★ line.
- If two-line TTLTRG★ groups are used, modules connect to defined TTLTRG★ group pairs (that is, lines 0 and 1, lines 2 and 3, lines 4 and 5, lines 6 and 7).
- If four-line TTLTRG★ groups are used, modules connect to defined four-line TTLTRG★ groups (that is, lines 0, 1, 2 and 3, lines 4, 5, 6 and 7).

### TTLTRG* line allocation protocols

Six standard allocation protocol procedures for TTLTRG★ lines are defined as follows.

### Synchronous (SYNC) trigger protocol

TTLTRG★ SYNC trigger protocol is of a single-line broadcasted trigger (Figure 9.4), not requiring an acknowledgement from any acceptor. Maximum trigger repetition rate is 12.5 MHz. SYNC trigger pulses from any source should have a minimum time of 30 ns and must not re-occur within 50 ns, although trigger pulses of 10 ns or more following a 10 ns or more period without triggers are accepted as SYNC triggers.

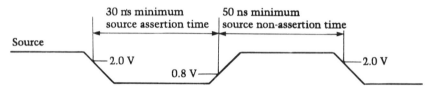

**Figure 9.4** *VXIbus TTLTRG* synchronous (SYNC) trigger protocol*

### Semi-synchronous (SEMI-SYNC) trigger protocol

TTLTRG★ SEMI-SYNC trigger protocol is of a single-line broadcasted, multiple-acceptor handshake (Figure 9.5). A source asserts the line low for a minimum time of 50 ns. To acknowledge the trigger, acceptors also assert the line low within 40 ns, holding it low if need be until they are ready for the next operation. Acknowledgement is thus completed when all acceptors have unasserted the line. Only after a further 50 ns may the source reassert the trigger, although in practice trigger pulses of 10 ns or more

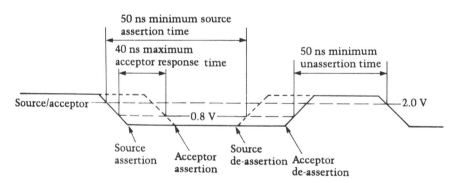

**Figure 9.5** *VXIbus TTLTRG* semi-synchronous (SEMI-SYNC) trigger protocol*

following a 10 ns or more period without triggers are accepted as SEMI-SYNC triggers. Operation is similar to GPIB data transfer using a talker's data valid (DAV) and a listener's not data accepted (NDAC) lines.

### Asynchronous (ASYNC) trigger protocol

This trigger has a two-line single-source, single-acceptor protocol, in which a source initiates an operation (Figure 9.6) by asserting the lower numbered TTLTRG* line of a defined pair for a minimum of 30 ns, while an acceptor acknowledges by asserting the higher numbered line for a minimum of 30 ns, although in practice trigger pulses of 10 ns or more following a 10 ns or more period without triggers are accepted as ASYNC triggers by source or acceptor.

This provides a useful means of handshake triggering between a VXIbus system and external instruments or between separate VXIbus systems.

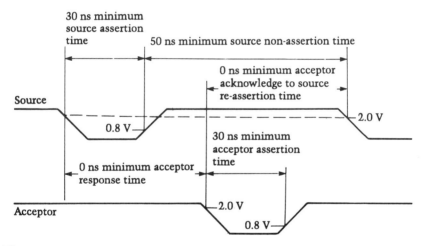

**Figure 9.6** *VXIbus TTLTRG\* asynchronous (ASYNC) trigger protocol*

### Clock transmission protocol

A TTLTRG* line may be used for clock signal transmission of 0 Hz to 12.5 MHz. However, in a fully loaded VXIbus subsystem rise time of TTLTRG* lines approaches 40 ns, so falling edges are always used for triggering.

### Data transmission protocol

TTLTRG* lines may be grouped together to transmit data in parallel. One line in the group is used as a clock, while data is synchronized to either its rising or its falling edge, or both.

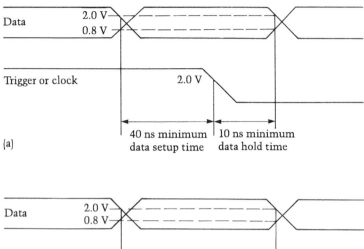

**Figure 9.7** *VXIbus TTLTRG\* data transmission protocol (a) using falling clock edge (b) using rising clock edge*

If a source uses a falling clock edge (Figure 9.7a) minimum data setup time is 40 ns while minimum data hold time is 10 ns. If a source uses a rising clock edge (Figure 9.7b) minimum data hold time increases to 40 ns. However, in practice acceptors accept any data having data setup and hold times of 7 ns or more.

### Start/stop (STST) protocol

This allows groups or clusters of modules to be synchronously operated. Slot 0 module drives any TTLTRG\* line and accepting modules respond at the next CLK10 rising edge (Figure 9.8). The time the driving line must be changed before CLK10 rising edge (START/STOP setup time) is a minimum of 50 ns, while the time it must be held after CLK10 rising edge (START/STOP hold time) is a minimum of 15 ns, although in practice any STST command with data setup and hold times of 7 ns or more will be accepted.

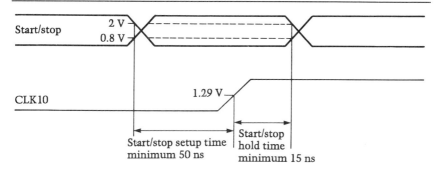

**Figure 9.8**  *VXIbus TTLTRG\* start/stop (STST) protocol*

### ECLTRG lines

ECLTRG lines are single-ended ECL lines, bused over the whole length of a VXIbus subsystem backplane. Lines are terminated at each end with $50\,\Omega$ terminations to $-2$ volts, which provide constant non-asserted low logic states on individual lines unless connected modules assert high states.

ECLTRG lines follow some basic rules:

- Maximum ECLTRG trace length from connectors on any module is less than 25 mm.
- Transmission line impedance between connector and source/acceptor circuit is $50\,\Omega$.
- Modules connected to ECLTRG lines use forward and reverse biased Schottky diodes between signal lines and an adequate transition bias voltage. Diodes should be as close to the receiver/driver as possible.
- If a module receives from, but does not drive ECLTRG lines it has either Schottky diodes as specified or a $150\,\Omega$ resistor in series with the transmission line connecting a connector's ECLTRG lines to the receiver.
- Modules may have only one driver/receiver load on any line.
- Modules must use drivers which provide a low output level less than the $-2$ volts termination supply.
- All lines become unasserted within one second or a required reset.
- Modules allocate ECLTRG lines to specific functions in groups of one or two.
- If single-line ECLTRG groups are used, modules connect to any ECLTRG line.
- If two-line ECLTRG groups are used, modules connect to defined ECLTRG group pairs (that is, lines 0 and 1, lines 2 and 3, lines 4 and 5).

### ECLTRG line allocation protocols

ECLTRG lines have functionally identical line allocation protocols to (albeit operating much faster than) TTLTRG★ lines, except for the ECLTRG start/stop protocol which is extended.

### Synchronous (SYNC) trigger protocol
SYNC trigger pulse should have a minimum time of 8 ns and must not re-occur within a further 8 ns, although in practice minimum times for both are 6 ns. Maximum repetition rate is 62.5 MHz.

### Semi-synchronous (SEMI-SYNC) trigger protocol
Here minimum source assertion time is 20 ns, maximum acceptor response is 15 ns, while the source must wait no longer than 5 ns to be able to re-assert the trigger. In practice, however, acceptors accept trigger of 13 ns or more following a 3 ns or more unassertion.

### Asynchronous (ASYNC) trigger protocol
Minimum source and acceptor assertion times are 8 ns (6 ns in practice), with an 8 ns period without triggers (6 ns in practice).

### Clock transmission protocol
If an ECLTRG line is used for clock purposes, timing requirements are as SYNC trigger protocol.

### Data transmission protocol
Minimum data set up time is 8 ns (6 ns in practice) while minimum data hold time is 4 ns (2 ns in practice).

### Extended start/stop (ESTST) protocol
ECLTRG ESTST protocol is a natural extension of the TTLTRG★ STST protocol, used when D–sized modules with connectors P3 are incorporated in a VXIbus subsystem. ECLTRG ESTST lines are used in addition to TTLTRG★ STST lines to synchronize modules with P3 connectors to those with only P2 connectors.

ESTST timing is synchronized to CLK100 (a 100 MHz clock signal which is synchronized with the 10 MHz CLK10 used by TTLTRG★ STST protocol), while a SYNC100 line is qualified by an ECLTRG ESTST line to initiate the ESTST operation. This means the original TTLTRG★ STST signal still determines the start/stop operation, with the benefit of a fixed time relationship between modules with P2 connectors and modules with P3 connectors. In effect, higher speed modules with P3 connectors can be tightly time coordinated with lower speed modules with only P2 connectors.

## Analog sumbus

This is an analog summing node, terminated by $50\,\Omega$ terminations to ground at each end of the VXIbus backplane. Also on the backplane is a protection network which clamps the sumbus to within $\pm 3\,$V for input currents of up to $\pm 520\,$mA, while not sourcing or sinking more than $\pm 1\,\mu$A or loading with more than 30 pF for voltages within $\pm 1\,$V.

Sumbus sources have equivalent output impedances of at least $10\,$k$\Omega$ in parallel with no more than 4 pF, and do not source more than $40\,$mA. When disabled, sources do not source more than $\pm 1\,\mu$A. Receivers have equivalent input impedances of at least $10\,$k$\Omega$ in parallel with no more than 3 pF, and usually high impedance buffer amplifiers are used. Sources and receivers must be within 25 mm of the relevant P2 connector pin (pin 32a).

As VXIbus sumbus is analog, it is susceptible to interference. Consequently backplane layout is designed to keep sumbus away from potential interference sources such as digital signals. To this end, the P2 connector pin used for sumbus is surrounded (pins 31a, 32a, 32b) by power supply voltages which act to screen sumbus.

## Power distribution bus

A maximum total of 268 watts is available from the VXIbus power distribution bus, over three connectors (P1, P2, P3) and in seven different regulated voltages. Table 9.7 lists voltages and maximum powers available at numbers of outlets on connector P1. Table 9.8 lists the same for connector P2, while Table 9.9 lists them for connector P3. Each power supply pin on any connector has a 1 A limit: it is this which effectively defines maximum power available, although power is naturally limited by mainframe capabilities.

Voltages of $+5\,$V, $+12\,$V and $-12\,$V exist on all connectors, and are intended for main power of modules, analog devices, disc drives and interfaces. A standby voltage of $+5\,$V is provided on connector P1, intended for use by devices which require battery backup in the event of

**Table 9.7** Voltages, numbers of outlets and powers available on the VXIbus power distribution on connector P1

| Voltages (V) | Number of outlets | Total power (W) |
|---|---|---|
| ground | 8 | |
| +5 | 3 | 15 |
| +12 | 1 | 12 |
| −12 | 1 | 12 |
| +5 standby | 1 | 5 |

**Table 9.8**   Voltages, numbers of outlets and powers available on the VXIbus power distribution bus on connector P2, plus total power available on connectors P1 and P2

| Voltages (V) | Number of outlets | Total power this connector (W) | Total power all connectors (W) |
|---|---|---|---|
| ground | 18 | | |
| +5 | 4 | 20 | 35 |
| +12 | | | 12 |
| -12 | | | 12 |
| +24 | 1 | 24 | 24 |
| -24 | 1 | 24 | 24 |
| -5.2 | 5 | 26 | 26 |
| -2 | 2 | 4 | 4 |

**Table 9.9**   Voltages, numbers of outlets and powers available on the VXIbus power distribution bus on connector P3, plus total power available on connectors P1, P2 and P3

| Voltages (V) | Number of outlets | Total power this connector (W) | Total power all connectors (W) |
|---|---|---|---|
| ground | 14 | | |
| +5 | 5 | 25 | 60 |
| +12 | 1 | 12 | 24 |
| -12 | 1 | 12 | 24 |
| +24 | 1 | 24 | 48 |
| -24 | 1 | 24 | 48 |
| -5.2 | 5 | 26 | 52 |
| -2 | 4 | 8 | 12 |

power failure. Voltages of +24 V and −24 V on connectors P2 and P3 are intended for analog source circuits and can be changed by on-board regulators or DC-to-DC converters, while voltages of −5.2 V and −2 V are for power and termination of high-speed ECL circuits. Each additional connector effectively adds extra power, up to either the maximum imposed by the mainframe power supply, or maximum limits imposed according to Tables 9.7 to 9.9. VXI power supply is discussed later.

### Clock and synchronization bus

VXIbus clock and synchronization bus comprises three pairs of signals:

- CLK10+ and CLK10−; a 10 MHz clock located on connector P2 (pins

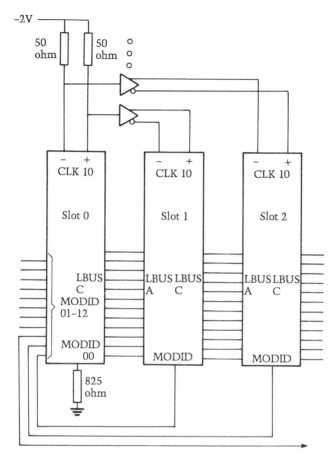

**Figure 9.9** *VXIbus backplane routings for clock, synchronization, module identification and local bus signals*

1c and 2c), with an accuracy no worse than 0.01% and a duty cycle of 50% ±5%.

- CLK100+ and CLK100−; a 100 MHz clock located on connector P3 (pins 31a and 32a), with an accuracy no worse than 0.01% and a duty cycle of 50% ±5%.
- SYNC100+ and SYNC100−; a synchronization signal located on connector P3 (pins 31c and 32c), with at least 4 ns setup time before, and at least 3.5 ns hold time from a CLK100 rising edge.

All are differential ECL signals and are screened from interference which may cause timing jitter by locating power supply voltages on surrounding pins on the P2 (pins 1b, 2b and 3c) and P3 (pins 30a, 30b, 30c, 31b and 32b) connectors.

Signals are generated by slot 0 module (or from an external source through slot 0) and are buffered on the backplane individually to each of the other slots. Backplane lines cater for $50\,\Omega$ signals, and any module accessing a signal provides a $50\,\Omega$ termination, with no more than two equivalent ECL loads. Backplane buffers do not insert more than 8 ns delay between slot 0 and any module, and add no more than 2 ns timing skew between signals at any two slots.

Synchronization signal SYNC100 is used to synchronize multiple modules with respect to a given CLK100 rising edge. Consequently a very tight time coordination between modules is possible.

Backplane signal routings for clock and synchronization signals (and module identification and local bus signals, see later) are shown in Figure 9.9.

### Module identification bus

VXIbus module identification bus (MODID) allows logical devices in a subsystem to be identified with a particular slot number, shown in Figure 9.10. Slot 0 module is connected to each other module, using the module identification bus and connector P2. One line (MODID01 to MODID12) connects to each module through a $16.9\,k\Omega$ resistor to $+5\,V$ on slot 0 module. Each module has an $825\,\Omega$ resistor to ground connected to its MODID line. Thus whenever a module is in position in a slot, a circuit is made (even if a module fails) which slot 0 module can detect.

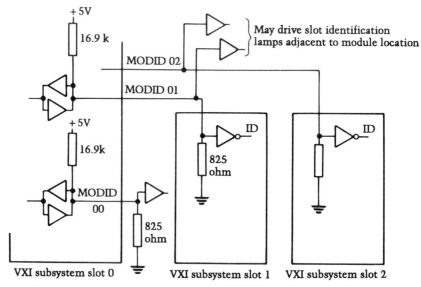

**Figure 9.10** *Module identification lines of VXIbus backplane*

The VXIbus backplane incorporates an 825 Ω resistor, too, which allows slot 0 module to check it is correctly placed in slot 0. An extra line in the module identification bus (MODID00) is connected to the backplane resistor through connector P2 (pin 30a). If MODID00 is low, slot 0 module is in slot 0; if high it is in another slot.

Effectively, the VXIbus module identification bus provides a means of automatic configuration: wherever modules are located in the subsystem, polling of MODID lines by slot 0 module can identify the location of modules. Module function can be pre-determined using its configuration registers (see later).

### Local bus

VXIbus local bus is a private bus, for use purely between adjacent modules. It is daisy-chained between modules (from slot 0 to slot 1, from slot 1 to slot 2, and so on), with the main purpose of allowing inter-module connections without need of physical wire jumpers across module front panels.

Connections are effected through the VXIbus backplane, and connectors (with the exceptions of slot 0 and slot 12 connectors) have 72 local bus lines, partitioned into 36 lines on each side (12 on each side, a and b, of connector P2, 24 on each side of connector P3 – actually some connections on connector P3 are on row b, too). Local bus lines on row a of connectors are labelled LBUSA, while those on row c are labelled LBUSC. Lines are numbered LBUSA00 to LBUSA35, and LBUSC00 to LBUDC35.

The VXIbus backplane connects LBUSC local bus lines on the connectors of one slot to LBUSA local bus lines on the connectors of the next high-numbered slot, with the exceptions of slot 0 and slot 12 (at each end of the VXIbus backplane).

A number of signal classes may be transmitted over VXIbus local bus, listed in Table 9.10, together with voltage and current limits. There is a possibility adjacent modules may transmit and receive incompatible classes of signals to each other, so a mechanical keying arrangement is used on

**Table 9.10** VXIbus local bus signal classes, together with voltage and current limits

| Signal class | Voltage (V) | Current (mA) |
|---|---|---|
| TTL | −5 to +5.5 | 200 |
| ECL | −5.46 to 0 | 50 |
| analog low | −5.5 to +5.5 | 500 |
| analog medium | −16 to +16 | 500 |
| analog high | −42 to +42 | 500 |

modules to indicate acceptable signal classes. However, the keying mechanism merely offers an *indication* and damage may still occur if adjacent module signal classes are incompatible.

### Star bus

VXIbus star bus comprises two sets of lines (STARX and STARY) from slot 0 module to each other module in a subsystem, for asynchronous communications purposes between modules. Lines are differential (STAR+ and STARX−, STARY+ and STARY−) and bidirectional.

Figure 9.11 shows how slot 0 module may be viewed at the centre of starbus, connected to each of twelve possible other modules in a subsystem. Starbus connections are made to slot 1 to 12 modules on connector P3 of VXIbus, on pins 28a and 29a (STARX+ and STARX−) and 28c and 29c (STARY+ and STARY−). Connections to slot 0 module, on the other hand, are individual STARX and STARY lines from each other module, located between pins 9a and 26c (STARX01+, STARX01−, to STARY12+, STARY12−) of the slot 0 P3 connector.

Starbus features high quality lines, similar in performance to clock and synchronization bus lines, with maximum timing skew of 2 ns between any two signals and an absolute delay of no more than 5 ns between slot 0 module and any other.

One of starbus's main uses is as a crosspoint switch, which selectively

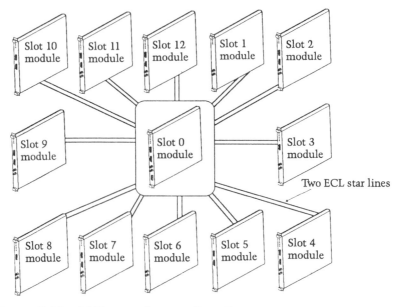

**Figure 9.11** *VXIbus starbus configuration*

routes signals from, say, a circuit under test, via any module to slot 0 module – which, in turn, routes signals to other modules. In this way a VXIbus system is tightly locked to the circuit's signals.

## Mechanical considerations

Mechanically, VXIbus is a mainframe chassis complete with card guides, connectors and backplane, enabling users to plug modules into slots in the mainframe chassis. Modules are standard sizes, listed earlier in Table 9.1. It

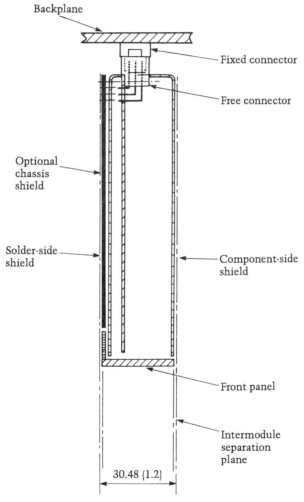

**Figure 9.12** *Top view of a shielded module in place in a VXIbus mainframe chassis*

is important to note that although modules are specific sizes listed, circuit boards used within modules may be smaller, though typically they are the same. System specification for VXIbus gives many rules relating to module dimensions and details which ensure modules are uniform. Modules made by any manufacturer to the specification therefore fit any given mainframe chassis. Modules may be thicker than the sizes listed, in multiples of given thicknesses, meaning multiple-width modules merely take up more slots. Where slots in a mainframe chassis are unused, blank filler panels are used as cover plates.

One advantage VXIbus has over VMEbus is its slightly greater module width allowing circuits to be screened with metallic shielding enclosures, which may be grounded to circuit or mainframe chassis ground. A top view of a shielded module housed in a mainframe chassis is shown in Figure 9.12. An optional, removable, mainframe chassis shield is allowed, and shields of adjacent modules may be in electrical contact (in areas near fron panels) using compressible gaskets or clips. Where modules are not grounded to chassis ground any adjacent chassis shield is insulated from ground. Where modules are grounded to chassis ground, on the other hand, modules and adjacent chassis shields may be in electrical contact.

Cooling of modules is undertaken by forced airflow through the mainframe chassis, air entering through vents in the bottom and leaving through top vents (Figure 9.13). The VXIbus specification makes no recommendation for cooling performance, instead leaving the onus on manufacturers of modules to determine cooling requirements for any

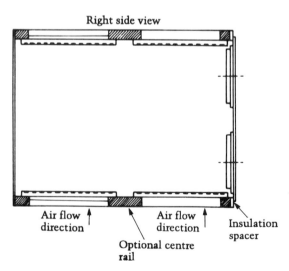

**Figure 9.13** *Airflow through a VXIbus mainframe chassis*

given module. It is recommended that modules are labelled with cooling specifications. System designers, using modules adequately labelled, then determine cooling requirements for the complete mainframe chassis.

Backplanes consist of a single circuit board, onto which module connectors are permanently attached and connections from all connector pins are made.

## Power supplies

The VXIbus backplane distributes a number of direct current power supply voltages around a subsystem:

- $+5$ V; main power source for most VXIbus systems, required by all systems for bus communications.
- $\pm 12$ V; for analog devices, disk drives, communications interfaces and so on.
- $\pm 24$ V; for analog signal sources, or derivation of other voltages using on-module regulators.
- $-5.2$ V; for ECL devices.
- $-2$ V; termination of ECL loads.
- $+5$ V standby; battery back-up to sustain memory, clocks and so on, if $+5$ V voltage is lost.

Voltages, along with variations, ripple caused by load, and induced ripple are listed in Table 9.11. Power limit is imposed by a 1 A maximum current per connector pin. Total power available at each connector is discussed earlier (see Tables 9.7 to 9.9).

**Table 9.11** Voltages, maximum ripple caused by load, and maximum induced ripple in a VXIbus subsystem

| Voltage (V) | Variation (V) | DC load ripple maximum (mV) | Induced ripple maximum (mV) |
|---|---|---|---|
| $+5$ | $+0.25$ to $-0.125$ | 50 | 50 |
| $+12$ | $+0.6$ to $-0.36$ | 50 | 50 |
| $-12$ | $-0.6$ to $+0.36$ | 50 | 50 |
| $+24$ | $+1.2$ to $-0.72$ | 150 | 150 |
| $-24$ | $-1.2$ to $+0.72$ | 150 | 150 |
| $-5.2$ | $-0.26$ to $+0.156$ | 50 | 50 |
| $-2$ | $-0.1$ to $+0.1$ | 50 | 50 |
| $+5$ standby | $+0.25$ to $-0.125$ | 50 | 50 |

## Devices and communications

Communications needs of all devices within a VXIbus system are catered for with a layered set of communications protocols, illustrated in Figure 9.14.

Each device has a set of **configuration registers**, containing information relating to the device's logical address, address space and memory requirements, type, model and manufacturer, (Figure 9.15). These configuration registers guarantee automatic system and memory configuration for each device, on application of power, at least to a minimum degree of communication. Devices with this minimum configuration level alone are known as **register–based devices**.

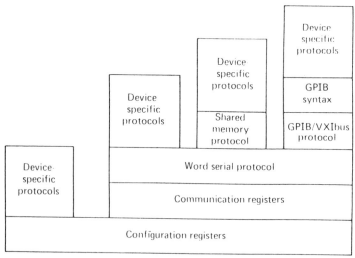

VXIbus communication layers

**Figure 9.14** *VXIbus communication protocol layers*

Register-based device

| | |
|---|---|
| $06_{16}$ | Offset register |
| $04_{16}$ | Status/control register |
| $02_{16}$ | Device type |
| $00_{16}$ | ID/logical address register |

**Figure 9.15** *Register-based VXIbus device configuration register*

A further set of registers, known as **communication registers** (Figure 9.16), is included if a device is to communicate at a higher level with other devices. Devices with communication registers are known as **message-based devices** and are all able to communicate, initially at least, with a specific protocol called **word serial protocol**. Once communication is established using word serial protocol devices may opt to progress to higher performance protocols. By no means a coincidence, VXIbus word serial protocol is very similar to GPIB communications protocol. One benefit of agreement by manufacturers to build devices which use standardized communications protocols such as word serial protocol is that devices from any manufacturer will be compatible. Higher level protocols may be defined later, on agreement by manufacturers, perhaps to be standardized into the VXIbus system specification.

Devices may be of purely memory form, say, RAM or ROM. Configuration registers of such **memory devices** are illustrated in Figure 9.17.

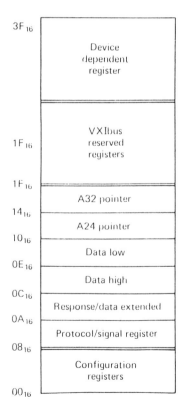

**Figure 9.16** *Message-based VXIbus device communication registers*

Memory devices

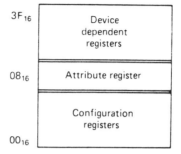

**Figure 9.17** *VXIbus memory device configuration register*

In some systems, **extended devices** are used which allow for definition of sub-classes of device types not defined in the original specification. Configuration registers of extended devices are illustrated in Figure 9.18, where the sub-class register allows for definition of both standard and manufacturer-specific extended device sub-classes.

These make up the four classes of VXIbus devices. In addition, there are two classes of VMEbus device which may be used in a VXIbus system. First, is the **hybrid device**; a VMEbus compatible device which has the ability to communicate with VXIbus devices, but does not comply with VXIbus device requirements. Second, is the **non-VXIbus device**; a VMEbus device not using or complying with any VXIbus requirement.

VXIbus devices of all classes fall into a hierarchical structure, illustrated in Figure 9.19.

Extended devices

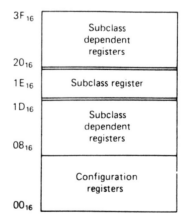

**Figure 9.18** *VXIbus extended device configuration register*

**Device classification**

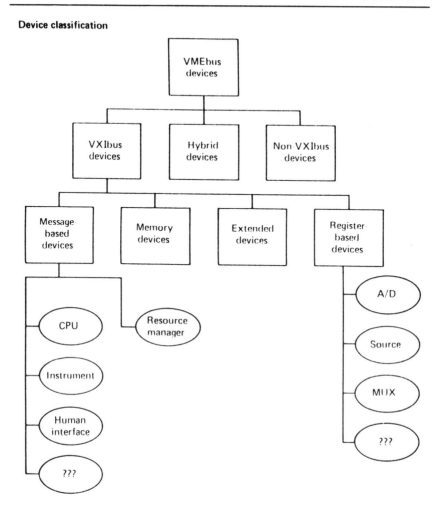

**Figure 9.19** *VXIbus device hierarchical classifications*

## Control

Control of the logical resources (memory, diagnostic procedures, self-test analysis and so on) of a VXIbus system is exercised by a **resource manager** device, often located in slot 0 and given the logical address 0. Resource manager functions are often coupled with general-purpose devices such as a VXIbus-to-GPIB interface, printer output and general timing functions. When positioned in slot 0, a resource manager is sometimes known as an **embedded controller**.

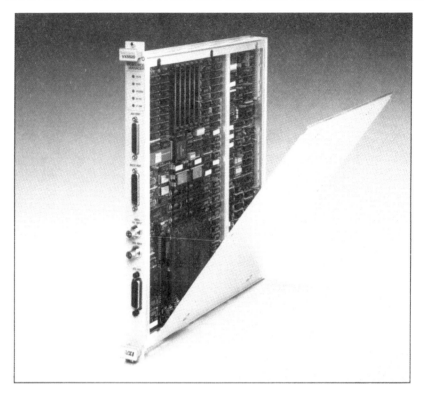

**Photo 9.4**   *VXIbus slot 0 resource manager (Tektronix)*

On application of power it is the resource manager's function to read the logical addresses of each device in the system, thereby locating all device configuration registers. Once these have been read the resource manager knows total system requirements, and allocates resources to suit.

Test instruments in any subsystem may be individual, in effect a single device, or may be connected together through the VXIbus backplane to become a **super-instrument**, comprising many devices. Super-instruments are logical – a module which forms part of a super-instrument during one test may be then logically separated, operating later either individually or in another super-instrument during another test.

Where systems feature more than one microprocessor-controlled device (typically in such super-instruments) the resource manager also determines hierarchies between them, allocating devices as **commanders** or **servants**. Commanders are message-based devices, which control one or more servants. Servants, however, may also be commanders and have servants of their own (Figure 9.20).

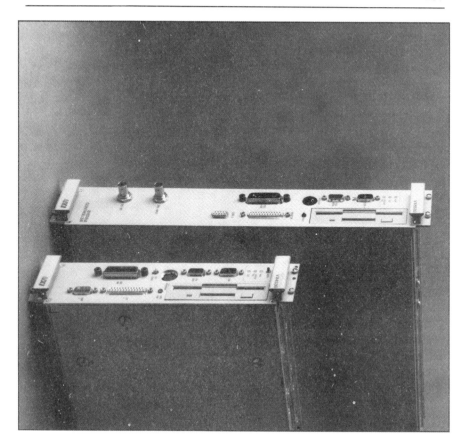

**Photo 9.5**   *VXIbus D- and C-sized embedded system controllers (Tektronix)*

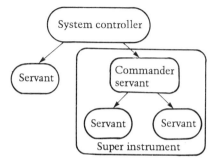

**Figure 9.20**   *Hierarchical structure typically possible on VXIbus, using devices acting as commanders, servants and super-instruments*

Readers may wish to consider **VXIbus** in greater depth than is possible here. If so, refer to:

- Report 88-0529R: *The VXIbus* by ERA Technology Ltd, Cleeve Road, Leatherhead, Surrey KT22 7SA. This gives a concise description of VXIbus, together with general descriptions of equipment and instruments available.
- *VXIbus system specification*, generally available from VXIbus device manufacturers. This is the in-depth specification of the VXIbus, detailing signals and timing considerations.

# Glossary

**Adaptive homing sequence**  synchronizing sequence applied to a circuit to ensure it is in a known state, automatically adapting to circuit behaviour

**Address command**  general-purpose interface bus command issued to specific devices

**Address-only cycle**  type of VMEbus data transfer bus cycle

**Admittance measurement**  see voltage forcing

**Analog sumbus**  VXIbus bus used as an analog summing node

**Arbiter**  a VMEbus functional module

**Arbitration cycle**  VMEbus process cycle allocating data transfer bus to modules requesting it

**Auto-delay**  test method which takes measurements at a number of intervals, comparing values with previous measurements until a predefined limit of difference is obtained. Synonymous with *auto-dwell*

**Auto-dwell**  see auto-delay

**Automatic test equipment**  a combination of one or more programmable test instruments, controlled by a physically separate computer (although all parts may be situated in one housing)

**Automatic test instrument**  test instrument with an internal computer, usually microprocessor-based, allowing automatic control of certain functions

**Backdriving**  test method used in overdriving, forcing an output's low output to a high state

**Bare-board**  a printed circuit board before component assembly

**Bed-of-nails fixture**  fixture comprising many spring-loaded test point probes onto which a tested printed circuit board is located. Thus, internal circuit nodes are accessed

**Bit pattern measurement**  application of test pattern vectors to a circuit, and later measurement of test pattern vector results

**Block transfer cycle**  type of VMEbus data transfer bus cycle

**Board under test (BUT)**   term often used to describe something tested by an automatic test equipment system

**Boundary scan**   test method linking scan registers of all scan test components of a product, such that all data held in the registers may be serially shifted out and observed

**Built-in self-test (BIST)**   test method using components with internal test sequences and procedures

**Bus-cycle emulation**   test method in which an automatic test equipment system assumes control of a tested product's bus. Synonymous with *bus-timing emulation*

**Bus timer**   a VMEbus functional module

**Bus-timing emulation**   see bus-cycle emulation

**Capture mode**   operation of a tested product when its memory is emulated by an automatic test equipment system. See memory emulation

**Clock and synchronization bus**   VXIbus bus carrying system clock and synchronization signals

**Cluster**   an isolated part of a circuit. Synonymous with *partition*

**Cluster test**   test procedure involving isolation of a part of a tested circuit. Synonymous with *partitioning*

**Combination test**   test procedure using both in-circuit and functional test methods

**Commander**   message-based VXIbus device, capable of controlling other devices

**Communication register**   VXIbus register allowing a device to communicate with word serial protocol

**Configuration register**   VXIbus register holding information relating to a device's logical address, address space and memory requirements, type, model and manufacturer

**Controller**   controlling device on general-purpose interface bus

**Current forcing**   type of analog test circuit. Synonymous with *impedance measurement*

**Data transfer bus**   bus used for transfer of binary data over VMEbus

**Data transfer bus cycle**   VMEbus process cycle defining data transfer over the data transfer bus

**Design for testability (DFT)**   inclusion of features at design stage of a product to allow easier testing after production

**Detent**   crimp in the side wall of a receptacle, which holds a spring contact test probe in position within the receptacle

**Device-dependent messages**   general-purpose interface bus messages passed between device functions of different devices, via interface functions

**Device functions**   functions of devices on general-purpose interface bus which are not dependent on the bus itself

**Device under test (DUT)**   term often used to describe something tested by an automatic test equipment system

**Diagnosis**   process defining what is wrong with a circuit, part of a circuit, or component

**Diagnostic resolution**   ability of an automatic test equipment system to identify individual faults

**Diagnostic test routines (DTR)**   type of target routine allowing a product's functional blocks to be tested while in memory emulation test procedure

**Digital guarding**   process using overdriving, to stabilize parts of a circuit to a known state

**DTB arbitration bus**   VMEbus bus used to ensure only one module control the data transfer bus at one time

**Dual-chamber fixture**   test method comprising two identical fixtures, allowing location and removal of products while another product is being tested. Synonymous with *dual-fixture test*

**Dual-fixture test**   test method comprising two identical fixtures, allowing location and removal of products while another product is being tested. Synonymous with *dual-chamber fixture*

**Embedded controller**   VXIbus controller with integral resource manager functions, in slot 0 of a VXIbus system

**Exhaustive test**   test method on digital circuits which tests all possible states of the circuit

**Extended device**   VXIbus device of a type not directly covered by the VXIbus specification

**Fixture**   that part of an automatic test equipment system which allows required contact between a tester and the product being tested

**Force node**   node of an analog guarding circuit

**Formatted output**   digital driver output which is manipulated to be of specific shape

**Four-wire testability bus**   bus formed by linking together four common inputs and outputs of all scan test components

**Functional module**   logical part of a VMEbus system

**Functional test**   test procedure simulating normal operation of products

**General-purpose interface bus (GPIB)**   data and control bus connecting automatic test instruments and a controller, according to IEC625 or IEEE488 standards

**Golden image**   term to describe an ideal product

**Go/nogo test**   test procedure based on functional test, which rapidly allows a product to be identified as totally working or not working. Individual faults are not identified

**Guarding circuit**   analog circuit to minimize effect of peripheral components around a measured component, nulling current through those components by equating voltages at all relevant points

**Guard node**   node of an analog guarding circuit

**Homing sequence**   synchronizing sequence applied to a tested circuit to initialize the circuit to a known state

**Hybrid device**   VMEbus device on a VXIbus system, with ability to communicate with VXIbus devices but not complying with VXIbus specification

**JACK★ daisy chain driver**   a VMEbus functional module

**Idle routine**   type of target routine, which keeps a product's microprocessor in a known condition during memory emulation test procedure

**IEC625**   standard defining an automatic test equipment system interface bus. See *general-purpose interface bus*

**IEEE488**   standard defining an automatic test equipment system interface bus. See *general-purpose interface bus*

**Impedance measurement**   see current forcing

**In-circuit test (ICT)**   test procedure in which products (printed circuit boards) are tested by allowing access to internal nodal points. Tests are then carried out on individual components or parts of products

**Instrument-on-a-card (IAC)**   test instrument located on a plug-in module

**Interface functions**   general-purpose interface bus functions which allow devices to receive, process or send messages. These are fixed and independent of device

**Interface commands**   special commands, issued by a controller on general-purpose interface bus

**Interface messages**   general-purpose interface bus messages passed between interface functions of different devices

**Internal scan test**   test undertaken on individual scan test components within a product

**Interrupt acknowledge cycle**   VMEbus process cycle initiated by interrupt handlers in response to an interrupt

**Interrupt handler**   a VMEbus functional module

**Interrupter**   a VMEbus functional module

**Listener**   device on general-purpose interface bus which can only receive data

**Local bus**   VXIbus bus used for private daisy-chained communications between adjacent modules

**Local messages**   general-purpose interface bus messages passing between device functions and interface functions of single devices

**Location monitor**   a VMEbus functional module

**Listener/talker**   device on general-purpose interface bus which can receive or send data

**Manufacturing defects analysis**   in-circuit test procedure, in which no power is applied to tested products. Synonymous with *pre-screening*

**Master**   a VMEbus functional module

**Measurand**   quantity being measured by a test or instrumentation system

**Memory device**   VXIbus device of purely memory form

**Memory emulation**   test method in which an automatic test equipment system memory is used as a tested product's memory

**Message-based device**   VXIbus device with communication register

**Modular automatic test equipment system**   test system comprising a computer controller along with peripheral automatic test instruments, located as modules within a single housing

**Module identification bus**   VXIbus bus identifying individual modules with a particular slot number

**Moving probe fixture**   fixture comprising two or more robotically-controlled arms which move over a tested printed circuit board. At the ends of the arms are test point probes, which access internal circuit nodes of the board

**Multiline messages**   general-purpose interface bus message passed over a group of signal lines

**Multimode test**   test procedure in which partition sizes are adapted by the automatic test equipment system in the light of results of previous tests. Synonymous with *polyfunctional test* and *variable in-circuit partitioning*

**Multiplexing**   test method allowing a test input or output to be connected to more than one node within a circuit

**Multiplexing ratio**   ratio of circuit nodes which may be accessed by a test system to the number of actual test inputs and outputs

**Nodeforcing**   test method used in overdriving, forcing a high output to a low state

**Non-VXIbus device**   VMEbus device on a VXIbus system, not using or complying with any VXIbus requirement

**Open architecture**   microprocessor-based system, independent of any particular microprocessor

**Overdriving**   test method allowing electrical isolation of parts of a digital circuit, by forcing previous stage outputs to known logic states. See *backdriving* and *nodeforcing*

**Parallel poll**   general-purpose interface bus poll sequence in which controllers check device operating conditions periodically

**Partition**   an isolated part of a circuit. Synonymous with *cluster*

**Partitioning**   test method which isolates and test parts of a circuit. Synonymous with *cluster test*

**Performance test**   see bus-cycle emulation

**Poll**   general-purpose interface bus sequence allowing devices to be serviced by a controller

**Polyfunctional test**   test procedure in which partition sizes are adapted by the automatic test equipment system in the light of results of previous tests. Synonymous with *multimode test* and *variable in-circuit partitioning*

**Power distribution bus**   VXIbus distributing power supply voltages to all modules

**Power monitor**   a VMEbus functional module

**Pre-screening**   in-circuit test procedure, in which no power is applied to tested products. Synonymous with *manufacturing defects analysis*

**Prioritized arbiter**   type of VMEbus arbitration cycle arbiter

**Priority interrupt bus**   VMEbus bus giving up to seven levels of interrupt

**Purchasor assessment**   assessment procedure in which a buyer tests products at the goods inward stage

**Quadrature detector**   see true-phase four quadrant measurement

**Rack and stack automatic test equipment system**   test system comprising a controlling computer, with peripheral automatic test instruments

**Read-modify-write cycle**   type of VMEbus data transfer bus cycle

**Read/write cycle**   type of VMEbus data transfer bus cycle

**Receptacle**   housing for a spring contact test probe, usually permanently positioned in a fixture and wired to the test system

**Register-based device**   VXIbus device with configuration registers

**Requester**   a VMEbus functional module

**Resource manager**   VXIbus device, usually in slot 0, which control logical resources of a system

**Round-robin-select arbiter**   type of VMEbus arbitration cycle arbiter

**Scan register**   internal parallel in, serial out register of a scan test component

**Scan test**   test method using scan test components

**Scan test component**   component with an internal scan register, which allows component inputs and outputs to be serially shifted out of the scan register on command

**Screen**   use of a serial test strategy, where first tester (the screener, or pre-screener) checks for common faults which can be simply tested

**Secondary address**   generous-purpose interface bus address given over more than one message

**Secondary message**   general-purpose interface bus message issued over more than one individual message

**Sense node**   node of an analog guarding circuit

**Serial clock driver**   a VMEbus functional module

**Serial poll**   general-purpose interface bus poll sequence in which devices request attention

**Servant**   VXIbus device controlled by another device

**Side bus**   private backplane bus between modules, separate from VMEbus buses

**Signature**   result of a digital test procedure, usually defining a correct response. If the response is not obtained the circuit is assumed to be faulty

**Signature analysis**   test method using signatures

**Single level arbiter**   type of VMEbus arbitration cycle arbiter

**Six-terminal measurement**   analog test circuit with six connections between tester and component. Synonymous with *six-wire measurement*

**Six-wire measurement**   see six-terminal measurement

**Slave**   a VMEbus functional module

**Spring contact test probe**   spring-loaded test point probe, typically used in bed-of-nails fixtures

**Spring rate**   force presented by the internal spring of a spring contact test probe against compression of plunger into barrel

**Star bus**   VXIbus bus arranged to provide asynchronous communications between modules, through the slot 0 module

**Status byte**   general-purpose interface bus message defining current operating condition to the controller

**Subsystem**   VXIbus housing, comprising up to 13 modules

**Super-instrument**   a logical test instrument comprising many devices in a VXIbus system

**Synchronizing sequence**   test pattern vector applied to a circuit to ensure the circuit is in a known state

**System clock driver**   a VMEbus functional module

**System under test (SUT)**   term often used to describe something tested by an automatic test equipment system

**Talker**   device on general-purpose interface bus which can only send data

**Target routine**   programs held in an automatic test equipment system's memory, used to control a tested product during memory emulation test procedure

**Testability**   ability and ease with which a product may be tested

**Test area management systems**   overall computer control of automatic test equipment systems and other systems within a manufacturing organization

**Test strategy**   plan and definition of an organization's test total requirements

**Three-terminal measurement**   analog test circuit with three connections between tester and component. Synonymous with *three-wire measurement*

**Three-wire measurement**   see three-terminal measurement

**Trigger bus**   VXIbus bus used primarily for intermodule communications

**True-phase four quadrant measurement**   test circuit in analog product test which allows measurement of parallel components. Synonymous with *quadrature detector*

**Turnkey automatic test equipment system**   complete test device, usually constructed to test just one product although may be adaptable to allow test of other products

**Two-terminal measurement**   analog test circuit with two connections between tester and component. Synonymous with *two-wire measurement*

**Two-wire measurement**   see two-terminal measurement

**Uniline message**   general-purpose interface bus message passed over a single signal line

**Universal command**   general-purpose interface bus command issued by a controller to all devices

**Utility bus**   VMEbus bus, used to maintain system operation, reset and timing

**Variable in-circuit partitioning (VIP)**   test procedure in which partition sizes are adapted by the automatic test equipment system in the light of results of previous tests. Synonymous with *multimode test* and *polyfunctional test*

**Vector**   digital test pattern of bits

**Vendor assessment**   assessment procedure in which a manufacture test products manufactured, thus buyers of products do not need to test them

**Virtual test channels**   use of scan register contents of scan test components surrounding a non-scan test cluster to access circuit nodes common to both cluster and scan test components

**VMEbus**   computer bus often used in automatic test equipment system architecture

**Voltage forcing**   test circuit in analog product test. Synonymous with *admittance measurement*

**VXIbus**   computer bus often used in automatic test equipment system architecture

**Walking out**   test method in sequential logic circuits, stepping test patterns though a circuit until they or their results are output

**Word serial protocol**   VXIbus message protocol, allowing message-based devices to communicate

**Working travel**   normally compression length of a spring contact test probe. Usually around two-thirds of total travel

# Worldwide addresses

## VMEbus and VXIbus

VMEbus International Trade Association
VITA Europe
PO Box 192
NL – 5300 AD Zaltbommel
The Netherlands

VMEbus International Trade Association
VITA USA
10229 N Scottsdale Road
Suite E
Scottsdale
AZ 85253

## STEbus

STE Manufacturers and Users Group
PO Box 149
Reading
RG6 3HB

The IEEE Computer Society
PO Box 80452
Worldway Postal Centre
Los Angeles
CA 90080

## Standards organizations and bodies

### American National Standards Institute (ANSI) and cooperating bodies

American National Standards Institute
1430 Broadway
New York
New York 10018

Electronic Industries Association (EIA)
2001 Eye Street Northwest
Washington DC 20006

Institute of Electrical and Electronics Engineers (IEEE)
345 East 47th Street
New York
New York 10017

### British Standards Institution

British Standards Institution (BSI)
Linford Wood
Milton Keynes
Buckinghamshire
MK14 6LE

### Deutshes Institut für Normung (DIN)

Deutshes Institut für Normung (DIN)
Burggrafenstrasse 4–10
D-1999 Berlin
30 T:26011

### International Electrotechnical Commission and National Committee

International Electrotechnical Commission
Rue de Varembe 3
CH-1211
Geneve 20
Switzerland

*Australia:*
Australian Electrotechnical Committee
Standard Association of Australia
80 Arthur Street
North Sydney
NSW 2060

*Canada:*
Standards Council of Canada
Communication Branch (Sales)
350 Sparks Street
Nbr 1200
Ottawa
Ontario
K1R 758

*Denmark:*
Dansk Elektroteknisk Komite
Strandgade 36 st
DK-1401
Kobenhavn K

*France:*
Comité Electrotechnique Français
12 place des Etats-Unis
F75783
Paris
CEDEX 16

*Germany:*
VDE-Verlag GmbH
Auslieferungsstelle
Merianstrasses 29
D-605
Offenbach aM

*India:*
Indian Standards Institution
Sales Service
Manak Bhavan
9 Bahadur Shah Zafar Marg
New Delhi 110002

*Italy:*
Comitato Ellettrotecnico Italiano
Viale Monza 259
1-20126 Milano

*Japan:*
Japanese Standards Association
1-24 Akasaka 4
Minato-Ku
Tokyo 107

*Netherlands:*
Nederlands Normalisatie-Instituut
Afd Verkoop en Informatie
Kalfjeslaan 2
2600 GB Delft

*Norway:*
Norsk Elektroteknisk Komite
Postboks 280 Skoyen
N – 0212
Oslo 2

*Spain:*
Comisión Permanente Española de Electricidad
Francisco Gervas No 3
Madrid 20

*Sweden:*
Standardiseringskommissionen i Sverige
Box 3295
S-10366 Stockholm 3

*Switzerland:*
Comité Electrotechnique Suisse
Association Suisse des Electriciens
Seefeldstrasse 301
8034 Zurich

*United Kingdom:*
British Standards Institution
BSI Sales Department
Linford Wood
Milton Keynes
Buckinghamshire
MK14 6LE

*United States of America:*
American National Standards Institute
1430 Broadway
New York
New York 10018

## Associated organizations

ERA Technology Ltd
Cleeve Road
Leatherhead
Surrey
KT22 7SA

## Magazines and periodicals

*Control and Instrumentation*
Morgan–Grampian
Morgan–Grampian House
30 Calderwood Street
Woolwich
London
SE18 6QH

*Electronic Engineering*
Morgan–Grampian
Morgan–Grampian House
30 Calderwood Street
Woolwich
London
SE18 6QH

*Electronic Production*
Angel Publishing
Kingsland House
361 City Road
London
EC1V 1LR

*Electronics Manufacture and Test*
Techpress Publishing Company
Northside House
69 Tweedy Road
Bromley
BR1 3WA

*Electronics Today International*
Argus Specialist Publications
Argus House
Boundary Way
Hemel Hempstead
HP2 7ST

*Printed Circuit Design*
PMSI
1790 Hembree Road
Alpharetta
Georgia GA 30201

*Printed Circuit Fabrication*
PMSI
1790 Hembree Road
Alpharetta
Georgia GA 30201

*Test*
Angel Publishing
Kingsland House
361 City Road
London
EC1V 1LR

# Further reading

Apart from numerous standards and publications mentioned throughout, the following publications and books should be of interest to readers. Dated publications indicate magazine articles.

Anon., Customised backplanes for ATE, *Electronic Production*, July 1990.

Anon., E-beam probing steps into the QA lab, *Electronic Engineering*, November 1988.

Anon., E-beam test station links design and test, *Electronic Engineering*, November 1987.

Anon., In-circuit tester offers 2048 discrete channels, *Electronic Engineering*, November 1986.

Anon., PCB verification, *Electronic Production*, July 1989.

Anon., Test instrument firms sign standards pact, *Electronics Weekly*, 25 April 1990

Bateson, John, *In-circuit Testing*, New York, Van Nostrand Reinhold.

Bonaria, Luciano, *The Quality Assurance System*, SPEA.

Bond, W. P., Automatic Test Equipment, a series of articles in *Electronics Today International*, October 1985 to February 1986.

Bradshaw, David, Bus wars, *Test*, September 1989.

Bradshaw, David, Does VXI make GPIB obsolete?, *Test*, September 1989.

Brindley, Keith, *Modern Electronic Test Equipment*, Oxford, Butterworth-Heinemann.

Brindley, Keith, *Newnes Electronics Assembly Handbook*, Oxford, Butterworth-Heinemann.

Buffoni, Dave, and Goringe, Barry, Keeping track of trends in backplanes, *Electronics Weekly*, 20 June 1990.

Caristi, Anthony, *IEEE488 General-purpose Interface Bus Manual*, London, Academic Press.

Clayton, P. Antony, *Handbook of Electronic Connectors*, Scotland: Electro-chemical Publications.

Coda, Spring Contact Test Probes, application and installation; probe and tool selection, Coda Systems Ltd., Braintree, Essex.

Colloms, Martin, *Computer Controlled Testing and Instrumentation. An introduction to the IEC625 IEEE488 bus*, Plymouth, Devon, Pentech Press.

Constantinou, Tat A., Statistical process control ... the other side of the coin, *I&CS*, November 1987.

Cordwell, Neil, Protection of integrated circuits during in-circuit test, *Electronic Engineering*, April 1988.

Costanzo, Lucia, Standard to unify test languages, *Electronics Times*, 26 April 1990.

Doumani, Alex, *VXIbus Higher Level Protocols*, Tektronix, Vancouver, WA.

Ellingham, Chris, Designing for test – someone else's problem, *Test*, November 1989.

Ellingham, Chris, Designing for test – the new challenges for system testing, *Test*, December 1989/January 1990.

ERA Technology, *Guide to Low-cost ATE*, ERA report: 89-0353, Leatherhead, Surrey, ERA Technology.

ERA Technology, *The VXIbus*, ERA report: 88-0529R, Leatherhead, Surrey, ERA Technology.

Evans, David, Designing for test – time for action, *Test*, July/August 1990.

Fisons, *Thermal Stress Screening*, Loughborough, Fisons Environmental Equipment.

Hagen, Michael S. *VXIbus System Configurations*, Tektronix Inc., Vancouver, WA.

Hamilton, Steve, Calculating the cost effectiveness of test strategies, *I&CS*, November 1987.

Hansen, Peter, A multimode programming strategy for VLSI boards, *Electronic Engineering*, February 1985.

Hansen, Peter, Coping with mixed technology boards, *Test*, July/August 1990.

Hansen, Peter, Strategies for testing VLSI boards using boundary scan, *Electronic Engineering*, November 1989.

Haworth, David A., *Hardware Overview*, Tektronix, Vancouver, WA.

Hay, Malcolm, A second generation performance tester, *Electronic Engineering*, November 1985.

Heath, Steve, *VMEbus User's Handbook*, Butterworth-Heinemann, UK 1989, CRC Press, US 1989.

King, Julia, ATE vs AOI, *Electronic Production*, April 1990

Lea, Tony, Testing SMT bare boards, *Electronic Production*, January 1990.

Marsh, David, VXIbus creates test flexibility, *Electronic Production*, July 1990.

Mawby, Terry, A primer on probes, *Electronic Production*, September 1989.

McCracken, A. Campbell, Bus-cycle emulation speeds test program generation and enhances diagnosis, *Electronic Engineering*, December 1989.

Parker, Kenneth P., *Integrating Design and Test: Using CAE Tools for ATE Programming*, Washington, Computer Society Press of the IEEE.

Penny, Andrew, Low cost ATE based on PCs, *Test*, September 1990

Peterson, Wade D., *The VMEbus Handbook*, Zaltbommel, The Netherlands, VMEbus International Trade Association.

Rittichier, Jeffrey, The changing rules of AOI, *Electronic Production*, October 1989.

Rohde & Schwarz, *Tester Compendium*, Munich, Rohde & Schwarz.

R S Components, STEbus computer boards and accessories, Data sheet 8919, R S Components, PO Box 99, Corby, Northants

Rudkin, A. M., *Electronic Test Equipment*, London, Granada Publishing.

Saunders, Glenn, Automating pcb test and repair, *Electronics Manufacture & Test*, March 1989.

Smith, Muriel and Cook, Steve, Techniques of SMD probing, *Electronic Engineering*, November 1986.

Society of Environmental Engineers, *Environmental Stress Screening*, SEECO, Herts.

Tektronix, *VXIbus System Specification*, Tektronix, Vancouver, WA.

Teradyne, *Teradyne Z8100 combinational test system*, Teradyne, Walnut Creek, CA.

Tooley, Michael, *Bus-based Industrial Process Control; Introducing the International STEbus Standard (IEEE1000)*, Oxford, Heinemann Professional.

Tooley, Michael, *Newnes Computer Engineer's Pocket Book*, Oxford, Butterworth-Heinemann.

Traister, John E., *Design Guidelines for Surface Mount Technology*, London, Academic Press.

Valdmanis, Janis, Advanced electro-optic sampling permits non-invasive testing of IC performance, *Electronic Engineering*, February 1989.

VMEbus International Trade Association, *The VMEbus Specification*, Zaltbommel, The Netherlands, VMEbus International Trade Association.

VMEbus International Trade Associations, *VMEbus Computer Applications*, quarterly publication for the VMEbus usergroup society, Zaltbommel, The Netherlands, VMEbus International Trade Association.

VMEbus International Trade Association, *64 bit VME, VME/Futurebus+ Extended Architecture*, Zaltbommel, The Netherlands, VMEbus International Trade Association.

Wakeling, Antony and McKeon, Alice, On automatic fault finding in analogue circuits, *Electronic Engineering*, November 1989.

Weber, Samuel, An HP test language reemerges as a standard, *Electronics*, May 1990.

# Index